I0069821

James Henry Emerton

Life on the Seashore or Animals of our Coasts and Bays

No. 1

James Henry Emerton

Life on the Seashore or Animals of our Coasts and Bays
No. 1

ISBN/EAN: 9783744716789

Printed in Europe, USA, Canada, Australia, Japan

Cover: Foto ©berggeist007 / pixelio.de

More available books at **www.hansebooks.com**

LIFE ON THE SEASHORE:

OR

ANIMALS OF OUR COASTS AND BAYS.

With illustrations and descriptions.

BY

JAMES H. EMERTON,

Author of Structure and Habits of Spiders.

NATURALISTS' HANDY SERIES.

NO. 1.

S. E. CASSINO & CO.,
BOSTON, MASS.

Printed at the Salem Press,
cor. Liberty and Derby Sts.,
SALEM, MASS.

PREFACE.

PREFACE.

In this book I have tried to give such explanations of some of the common animals of the New England coast as have often been asked for at the shore and in museums by persons little acquainted with zoology, and to give such directions about collecting and observing them as have been found useful to students, who come to the shore for a short time in the summer to study animals that they before knew only from pictures. For more complete accounts of the same animals the reader is referred to larger works, such as Gould's Invertebrata of Massachusetts, Verrill's Invertebrate Animals of Vineyard Sound, the Reports of the U. S. Fish Commission, and the special papers on particular groups of animals which are referred to in the general works just named. The common names of animals, like squids and lobsters, have been used as far as practicable, but for the majority of them the customary latinized names, by which they are known among naturalists have been given; and it is believed they will not be found more troublesome than the sea-peach, sea-corn, devil-fish, conger-eel, etc., which are used so indiscriminately by fishermen. Many of the figures have been drawn from nature for this book and the rest are copied from those which have appeared in the Reports of the U. S. Fish Commission, the Invertebrata of Massachusetts and other works; and I would here express my thanks to Prof. A. E. Verrill for the use of a large part of these figures and for other assistance, and also to Prof. S. I. Smith, Mr. A. Agassiz and the publishers of Tenney's Zoology.

TABLE OF CONTENTS.

TABLE OF CONTENTS.

	PAGE
ANIMALS BETWEEN TIDES	I
SHORE ANIMALS	II
Barnacles	12–14
Mussels	14–16
Snails	16, 17
Limpets	18, 19
Clams	19–22
Worms	23–29
Isopods	29, 30
Fiddler Crabs	31, 32
ANIMALS NEAR LOW–WATER MARK . . .	33
Lobsters	33–36
Shrimps	37
Crabs	38–44
Snails	45–49
Oysters	49–51
Scollops	51, 52
Teredo	52, 53
Worms	54, 55
Ascidians	56–59
Starfishes . , ...	59–62

TABLE OF CONTENTS.

PAGE

Sea-eggs 63–65
Sand dollar 66
Holothurians 66
Sea-anemones 67–70
Polyps 70–74
Hydroids 74, 75
Sponges 75–77
Infusoria 77, 78
Acinetæ 79

SURFACE ANIMALS 81
Copepods 83
Crustacea 84–86
Worms 86–88
Cuttlefishes 89–94
Pteropods 94, 95
Snails 95, 96
Ascidians 96, 97
Young starfishes and sea-eggs . . . 97
Young Lophothuria 98
Jelly-fishes 98–108

BOTTOM ANIMALS 109
Dredge-nets 109–111
Trawls 112, 113
Tangle 113, 114
Dredging 114–120
Clams 121
Laminaria 123

TABLE OF CONTENTS.

PAGE

Polyzoa 124

Chitons 125, 126

Red Crab 127

Spirorbis 127

Sea-anemone 127

Crustacea 128

Worms 121, 128–130

Snails 130

Holothurians 131

Ascidians 131

Shrimps 132

Amphipods 132, 133

Snails 122, 133, 134

Brachiopods 135

Polymastia 136

Polyps 136

Echinoderms 137

Index 141

LIST OF ILLUSTRATIONS.

FIG.		PAGE
102.	Acineta	78
160.	Alcyonium carneum, compound polyp	136
144.	Amphitrite ornata	118
140.	Angulus tener	114
16.	Anomia, under side of	19
111.	Architeuthis princeps, squid	93
68.	Ascidian, young	59
151.	Ascidians covered with sand	126
99.	Ascortis fragilis, ciliated cells of	76
155.	Astarte sulcata	130
78.	Asterias, starfish	65
93.	Astrangia Danæ, five polyps of	74
118.	Aurelia flavidula, common jelly-fish	99
119.	" " first stage of	99
106.	Autolytus	87
1.	Barnacle, young	12
1a.	" second stage of	12
2.	" anatomy of	13
137.	Boat dredging, diagram of	113
132.	Bolina aleta, jelly-fish	107
157.	Boltenia Bolteni, long-stalked ascidian	133
149.	Buccinum undatum, whelk	123
72.	Bugula turrita, enlarged twice	60
74.	" " branch of	62
43.	Callinectes hastatus, blue crab	42
40.	Cancer irroratus, common crab	37
47.	Caprella geometrica	46
113.	Cavolina tridentata	95

LIST OF ILLUSTRATIONS.

FIG.   PAGE

100. Chalina oculata, sponge   77

28. Cistenides Gouldii, brightly-colored worm   25

95. Clava leptostyla, polyp   75

112. Clione papillonacea, pteropod   94

27. Clymenella torquata, common beach worm   25

103. Copepod   83

159. Corymorpha pendula   35

96. Coryne mirabilis, jelly-fish, cluster of   76

96.   "   "   "   enlarged   76

125.   "   "   with tentacles enlarged   104

39. Crangon vulgaris   36

14. Crepidula fornicata, limpet, under side of shell   19

15.   "   plana, limpet, shell of   19

82. Cribrella, young   68

71. .Crisia eburnea   60

73.   "   "   branch of   61

141. Cyprina islandica   115

58. Dendronotus arborescens, snail   52

105. Diastylis quadrisponosus, Cumacea   86

143. Diopatra cuprea   117

60. Doris bifida, snail   53

134. Dredge   110

87. Echinarachnius parma, sand-dollar   70

84. Echinus, common sea-egg   69

87a.   "   young of   70

18. Ensatella Americana, razor clam   20

57. Eolis diversa, snail, young of   51

59.   "   pilata, snail   53

46. Eupagurus Bernhardus, hermit crab   45

77. Flustrella hispida, young polyzoa   64

33. Gammarus ornatus, amphipod   27

36. Gelasimus pugilator, fiddler crab   31

37. Homarus Americanus, lobster   34

128. Hybocodon prolifer, jelly-fish   105

FIG.		PAGE
146.	Hydractinia polyclina, female cluster	120
146.	" " male "	121
34.	Idotœa irrorata, isopod	29
133.	Idyia roseola, jelly-fish	107
55.	Ilyanassa obsoleta, snail	50
120.	Jelly-fish, early stage of	100
121.	" young, ready to swim away	100
56.	Lacuna vincta, brown-shelled snail	50
3.	Lapis fascicularis, floating barnacle	13
30.	Lepidonotus squamatus, scaly worm	26
32.	Leptoplana variabilis, planarian	27
88.	Leptosynapta Girardii, holothurian, upper end of	71
89.	" " plates and hooks from the skin of	71
45.	Libinia canaliculata, spider crab	44
35.	Limnoria lignorum, wood-bearing isopod	30
48.	Limulus polyphemus, horse-shoe crab	46
31.	Lineus viridis, nemertine	26
10.	Littorina litorea, periwinkle	17
12.	" palliata	18
11.	" rudis	18
38.	Lobsters, young	35
108.	Loligo pallida, squid	89
110.	" " pen of	92
109.	" Pealii, clusters of eggs of	91
161.	Lophothuria Fabricii, echinoderm	137
117.	Lophothurian, young	98
50.	Lunatia heros, snail	48
51.	" " crawling	48
53.	" " eggs of, on sand	49
52.	" " teeth of	49
23.	Macoma fusca, round clam	23
21.	Mactra solidissima, hen clam	22
154.	Margarita obscura	130
42.	Megalops, young crab	39

LIST OF ILLUSTRATIONS.

FIG.   PAGE

69. Membranipora pilosa, polyzoa, much enlarged 60

70.   "   "   single animal expanded 61

75.   "   "   young polyzoa 63

76.   "   "   swimming young, seen edgewise . . . 63

90. Metridium marginatum, common sea-anemone 72

6. Modiola modiolus, mussel 16

5.   "   plicatula, mussel 15

67. Molgula manhattensis, simple ascidian 59

17. Mya arenaria, common clam 20

104. Mysis stenolepis 85

4. Mytilus edulis, common mussel 14

24. Nereis virens, common bait worm, head and front segments of . . 23

26.   "   "   head of 24

25.   "   "   section of 24

129. Nanomia cara, compound jelly-fish 105

123. Obelia, jelly-fish 102

124.   "   commissuralis, branching hydroid 103

83. Ophiopholis aculeata, starfish 68

61. Oyster, young of 54

62.   "   "   55

63.   "   "   seen edgewise 56

64.   "   "   56

152. Pandalus annulicornis 127

42. Panopæus depressus, mud crab 41

122. Peachia parasitica, sea-anemone 101

65. Pecten irradians, scollop 57

150. Pentacta frondosa, holothurian 125

148. Phascolosoma cementarium, enlarged 122

130. Physalia arethusia, compound jelly-fish 106

44. Platyonichus ocellatus, lady crab 43

131. Pleurobrachia rhododactyla, jelly-fish 107

29. Polycirrus eximius, mud worm, tentacles extended . . . 25

7. Purpura lapillus, snail 16

8.   "   "   egg cases of, ... 17

49. Pycnogonidæ 47

FIG.   PAGE

92. Sagartia leucolena, white-armed sea anemone 73
116. Salpa Cabotti, ascidian 97
115. "   "   with chain of young within 96
19. Saxicava arctica, clam 21
91. Sea anemone, section of 73
85. Sea-egg, pedicellaria of 69
86. "   porous plate of 69
97. Sertularia pumila 76
138. Sieve, for separating contents of dredge 113
22. Siliqua costata 22
114. Snail, veliger of 96
81. Starfish, development of 67
79. "   spine of 69
81. "   young 66
156. Sternaspis fossor, worm 132

136. Tangle 112
13. Tectura testudinalis, limpet, shell of 18
158. Terebratulina septentrionalis 134
66. Teredo navalis, ship worm 58
94. Thammocnidia spectabilis 75
126. Tiaropsis diademata 104
127. Tima formosa 104
145. Trachydermon ruber 120
135. Trawl 111
54. Tritia trivittata, snail 50
142. Trophonia affinis, front segments of 116

153. Unciola irrorata 129
9. Urosalpinx cinerea, snail 17

20. Venus mercenaria, quahaug 21
101. Vorticella, from hydroid stems 78

107. Worm, young 88

139. Yoldia limatula 114

41. Zoea, young crab 38

ANIMALS BETWEEN TIDES.

ANIMALS

BETWEEN TIDES.

ONE of the most interesting things about the seashore to
the visitor from inland parts of the country is the difference
between its plants and animals, and those with which he is
familiar on the land or in fresh-water ponds and rivers. The
flowering plants, which cover with shrubs and weeds every
part of the dry land where they can get a foothold, are rep-
resented in the salt water by only one species, the eelgrass,
which grows in immense quantities on sand or mud just
below low water, and when the tide is out settles down into
a thick mat through which a boat can hardly be pushed. In
the summer the simple green flowers in a groove on the leaf-
like stem float out on the surface at low tide, but are covered
again when the water rises. Most of the marine plants belong to another class, the "algæ," some of which also grow
in fresh water. To the algæ belong the brown "rockweed"
and "Irish moss" that grow in slippery bunches over the

stones for some distance above low water, where they are left uncovered and partly dry for half the time, and the more delicate red and green seaweeds that live in deeper water.

Between land and marine animals the difference is quite as great. On land no animals are more common and of a greater number of kinds than the insects, while in the sea only about a dozen species live in their young condition near the shore. On the other hand, the crustacea, to which the lobster and crabs belong, are almost entirely marine, only the sow bugs, the crayfish and a few crabs living on land. Almost all the worms, too, live in the sea, the leeches and awkward earthworms giving no idea of the variety and beauty of this class of animals. Of the polypes and jelly fishes only the little hydra lives in fresh water, and none on land. The echinoderms, to which the starfishes and sea eggs belong, are all marine. The great number and variety of marine mollusks, compared with those of the land and fresh water, are well known by their shells, which form a large part of every natural history collection.

The readiest places to look for marine animals are along beaches near high water mark, among the sand and rubbish thrown up by the waves, but a large part of the objects found here are liable to be broken or worn by the water and to have the softer parts decayed. Just after a storm, however, a great deal may be found here, for animals which live usually below low tide are often loosened and thrown up fresh on the shore. Other objects of interest may be found among the roots of seaweeds which are often thrown up at such times in large quantities. If, however, one wants to see the animals of the shore to advantage, he must look farther down the beach where they are alive and in their natural condition. Everything here depends on the tides, for unless this is low very little can be reached. The almanac should be consulted beforehand, and, if possible, a time selected about the new or full moon, when the tides rise highest and also run out lowest. For collecting along the shore but little apparatus is needed; a few bottles with wide mouths in a basket or pail in which they will stand up, a trowel or strong knife for digging and prying, and a pair of small forceps, such as are made for dentists and jewellers, are the most useful things on a rocky shore or about wharves and bridges. Where there are pools of water left by the tide or wherever collecting can be done from a boat a dip net is wanted, and on sandy or muddy shores, a spade and sieve. A fish net is not fine enough for a naturalist's use, but one made of some kind of open cloth must be carried. The linen cloth sold for milk strainers, embroidery canvas, or thin towel linen, are good materials for the purpose. The net hoop should be of brass, as iron, even if tinned or covered with zinc, sooner or later becomes rusty and spoils the net. It is best to have the bottom of the net round so that there shall be no corners to catch dirt. Some collectors have long handles to their nets, so as to be able to reach objects at a distance or in deeper water, but for most purposes a stick three or feet long is sufficient, and can be carried and handled much more easily than a longer one.

In whatever place one goes about the salt water he is sure to get dirty, and should not wear anything that is intended to be worn in town afterward. Salt water spoils thin shoes,

and if the water is too cold or the shore too rough for bare
feet, rubber boots or still better stout leather ones are comfortable. The best places to find a large variety of marine
animals in a short time are rocky shores, or the posts of
wharves and bridges where these are in deep and tolerably
clear water. In such places one can work from a small
boat tied to the posts or held by a boat hook, and have his
pails and bottles within convenient reach within the trouble
of carrying them about with him. If two persons can go
together and one manage the boat while the other hunts, so
much the better, but do not try collecting with a boat full of
people if you can help it. After picking up enough of the
starfishes, barnacles and other things that are in plain sight,
gather a few bunches of mussels and hydroids to look over
at leisure and pick out the smaller animals among them ; but
do not take more than a handful or two, or delay examining
them, for most of the animals will die in a few hours if left
crowded together in a bucket, and so foul the water as to
kill the whole of them. Of the larger specimens only as
many should be taken as can be provided with clear water
and swimming room in the bottles at hand, and if the weather
is warm or the excursion extends over several hours, the
water should be frequently changed.

On a rocky shore, the best place to look is under stones,
the farther down the beach the better, and here again it is
an advantage for two persons to work together and so be
able to manage larger stones. A small iron bar is a help in
many cases to start up stones that have settled tightly together so that they can be turned over. A little practice

will show which stones are most likely to have something under them and they are usually those of moderate size and not too deeply settled into the sand.

If the shore is flat and smooth, many things can be found by wading just below low water and picking up with a dip net the crabs, mollusks, etc., that have been driven down by the receding tide. Where there are not too many rocks or patches of grass, a great many little fishes and shrimps can be caught by a fine seine which two persons carry out as far as they can wade and then draw as rapidly as possible up the beach. But the greater part of the inhabitants of sandy and muddy shores live under ground and can best be found by digging. Some species show where they are buried by a hole at the surface, or by a pile of sand coiled up like that near the hole of an earth worm, and others build tubes which extend a short distance above the sand. These tubes or holes extend far down into the sand, and the occupants are most of them ready to retreat downward at the least alarm; so that they must be dug quickly by one stroke of a long spade so as to cut off their retreat or their tails before they know what the matter is. A large part of the under-ground worms, however, show no signs of their presence on the surface and the sand must be dug up here and there until a good place is found. The larger clams and worms can be picked out easily enough, but there are many smaller ones so covered with dirt that they are likely to be missed; and to find these it is a good plan to put the mud into a sieve and shake it up and down in the nearest water, until the greater part of it is washed through, leaving a small amount of gravel, shells and worm tubes that can be easily picked over. A sieve will also be found useful for examining the mud and sand dredged in deeper water as described further on.

After bringing in the spoils of a collecting excursion, they need first of all to be sorted and put into larger vessels with clear water in which they will perhaps expand and show themselves. Dead or sick specimens should be removed and thrown away or put in alcohol for future study. Seaweeds and hydroids which have been brought ashore should then be floated out in pans of water and carefully searched for worms and mollusks that may live among them; and after looking them over once they should be left standing a few hours when other specimens will probably come into view. The next morning when the water has begun to foul, still other animals will be found loose about the dishes having found their tubes and holes uncomfortable during the night. The animals which are to be kept alive several days for observation should be put in a shady and cool place and the water changed often enough to keep it cool and clean; and when the change is made the slime should be washed off the dishes and any other dirt picked out. It is best not to try to feed them, it makes the water impure and they get along well enough without it, living on what they ate before being caught or may pick up in the water around them. A large aquarium can be kept in operation all the year without changing the water if a sufficient number of plants can be made to grow in it and the animals be not too numerous. For this purpose the green seaweeds are best and should be put in the tank long before it is to be used for animals. Small bright looking plants attached to stones are the best, and the new growth that attaches itself to the sides of the aquarium should be allowed to grow as far as it can without concealing its contents. As the plants become established, the dirt in the water decays rapidly and it becomes clearer and clearer, and then such animals as are found to be hardy may be put in, a few at a time. The aquarium should be in a cool place and be kept carefully clean, dead plants or animals and dirt of every kind being removed as soon as noticed. An aquarium can be kept in healthy condition for a long time without the use of plants, if the water is aerated by stirring or pouring out and in again several times a day.

As soon as a tide's collecting is looked over the need of a magnifying glass will be felt. One magnifying five or six times can be bought at any instrument dealer's, and will be very useful and all that is needed if one has little time to spare ; but a good microscope with compound glasses and stage and other conveniences will open to view a set of small objects that without it have to be passed by altogether. For use at the seashore a small and simple microscope is the best. A short one that can be used upright without straining the neck or using a high stool is more convenient than a long instrument even if the latter has some optical advantages. If an expensive instrument can be had let the money be put into the glasses rather than complicated stands and stage arrangements. Great care must be taken to keep the microscope clean, salt water should be wiped up as soon as spilled and when rust shows itself, as it surely will in the damp atmosphere of the seaside, it should be rubbed off with oil without the use of polishing powder of any kind as that would remove the lacker and make matters worse.

Knives and forceps and all kinds of metal tools should be washed in fresh water and dried as often as possible, and occasionally oiled and rubbed with emery or some softer powder if it will answer the purpose. The glasses of the microscope ought never to be touched with the fingers, or dirty cloth or paper. If they need wiping, as they occasionally will, do it with a clean handkerchief or a piece of washleather that is kept shut up out of the dust. When examining objects under the microscope use the lowest magnifying power that will show what is wanted. Living animals can be examined in watch glasses, which are the most convenient small dishes for many purposes. They are, however, unsteady on account of their rounded shape; and, unless the microscope has a movable stage, should always be placed on a strip of glass on which they can be slid about on the stage with less danger of spilling their contents. In examining animals in a watch glass or any open vessel another difficulty arises, owing to their constant motion, and it is usually better, and with high magnifying powers necessary, to put them between two pieces of glass with enough water to fill up the space between the glasses around them and make the whole appear clear. For this purpose a strip of thin plate or ordinary window glass of the customary size of one inch by three may be used with a smaller piece of very thin glass, such as is made specially for this use, laid over it. The glasses, especially the thin cover, should be wiped clean and not touched with the fingers except at the edges. The object should be placed as near the middle of the larger glass as possible, with a drop of water and the thin glass laid over it. If the object is thick or so soft as to be in danger of being crushed by the weight of the covering-glass, the cover may be propped up by bits of paper or wax so as to be level and just touch the object lightly. Care must be taken that no water gets on the top of the cover, and if it does the cover should be taken off and wiped or the water removed by soaking it up with a brush or a piece of paper or cloth.

The best way to pick up small floating animals is by a glass tube which is kept closed by the finger on the upper end until the lower end is close to the object, when the finger is removed and a sudden current of water runs up the tube carrying the specimen with it. For still smaller things and for taking up small quantities of water or other liquids, a common medicine dropper with a rubber bulb on one end is better.

For preserving small animals the most useful fluid is alcohol, such as is bought for burning and cleaning purposes. It is better for most objects to be put first into weak alcohol and after a day or two into stronger, as by this means they contract more gradually and evenly than if put into strong alcohol at once. It is often necessary to change the alcohol again after a few days, as the water which the specimen contains gradually dilutes the alcohol until it is too weak to prevent decay. Starfishes or crabs that are to be dried do better if kept a short time in alcohol beforehand so that the muscles may become hardened and dry without decaying.

SHORE ANIMALS.

SHORE ANIMALS.

THE strip of shore left uncovered twice a day by the changes of the tide is the home of a large number of animals some of which prefer to be thus exposed to the air part of the time.

Almost up to the limits of high water live the barnacles with white conical shells attached firmly to the stones and posts. Although they have shells the barnacles do not belong with the snails, but with the crabs, and in their young days swim about free in the water as many other young crabs do. Fig. 1 is a young barnacle such as may be caught in the spring swimming about at the surface of the sea. It has a three cornered shell, a single eye and three pairs of limbs. As they get older their shape changes to that of Fig. 1a and they become flattened sidewise and have a bivalve shell. They now have six pairs of swimming feet which extend beyond the edges of the shell and enable the young barnacle to swim rapidly through the water; at this stage

11

the antennæ become much larger and the animal uses them to hold itself to any stationary object upon which it may rest. This habit at length leads the barnacle to settle down

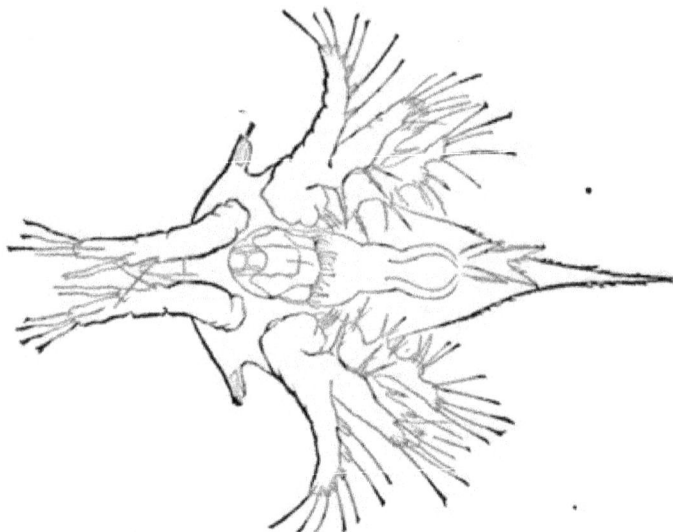

Fig. 1.—Young Barnacle enlarged fifty times.

for life, for the antennæ become attached to the object to which they hold and the animal finds itself fastened down

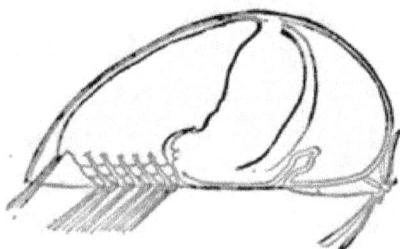

Fig. 1a.—Young Barnacle in the second stage ready to attach itself, enlarged twenty times. [From Darwin.]

by the head, Fig. 2. From the antennæ grows the flat base by which the barnacle is attached, or in some species a long stem as in the floating barnacle, Fig. 3. For some time after attachment the shape of the young barnacle remains to all outside appearances the same, but when it next moults it appears with a shell of a very different kind. It has two valves as before which now open upward, but below and around these are six smaller pieces that spread out to the stone on which it is fastened. They are at first soft like the old bivalve shell but soon become stiffer through deposits of lime. As the animal grows, the shell increases in size by additions to the edges

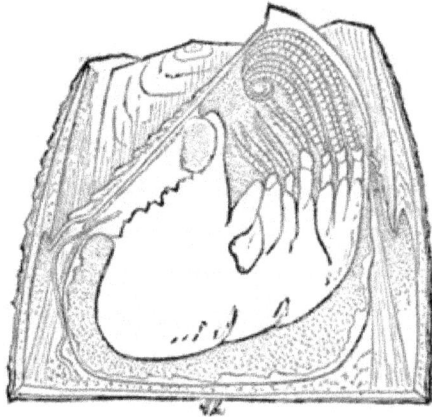

Fig. 2.—Barnacle with the shell of one side removed to show the softer parts. Above are the fringed appendages below which in the middle of the animal is the mouth. Under the middle of the base of the barnacle are the remains of the antennæ by which it attached itself. [From Darwin.]

of each piece; but the pieces around the sides grow faster than the movable valves and increase the height of the

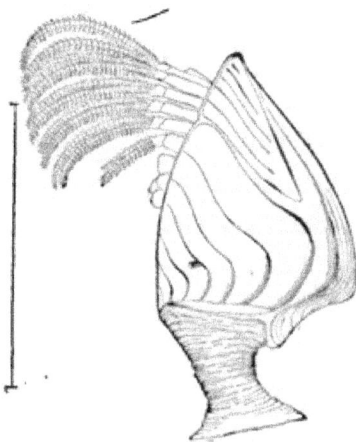

Fig. 3.

barnacle faster than its width. The skin of the softer parts of the body is moulted from time to time and these floating skins are among the most common objects found at the surface and are easily mistaken for jelly-fishes.

After the barnacle is fastened down by the head the appendages with which it used to swim become useful to keep up a current of water through the shell and wash in the smaller animals on which the barnacle lives. Each appen-

dage is divided into two branches covered with fine hairs
and they form all together a fan-shaped organ, which is
thrust out of the shell to its full extent and then closed
downward with a grasping motion toward the body. It is
then moved back between the valves, thrust out again, and
the motion repeated for hours together.

While the tide is out the barnacles close their shells and
in this position can live through several hours of the hottest
weather without becoming dried; but as soon as the water
rises high enough to cover them, they one by one put out
their scoops until the whole surface of the rock or post
seems to be in motion.
If some of them can
be detached without
breaking and put in a
dish of water they will
soon put out their
scoops and go through
their graceful motions
as well as on their native grounds.

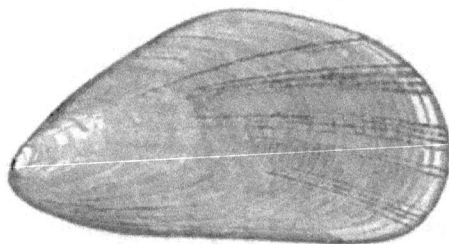

Fig. 4.—The Common Mussel, *Mytilus edulis*,
with the shell closed.

Almost as far up as the barnacles and down much lower,
the posts of wharves and bridges are often black with mussels,
Fig. 4, of all sizes, hanging together in clusters and furnishing hiding places for a great variety of worms and mollusks.
If a mussel be detached and put in a glass of water it soon
shows how it fastens itself down. It reaches out from the
shell a soft organ called the "foot" with which it creeps
slowly about, but as soon as it gets into a comfortable place
it presses the end of the foot against the dish and secretes
in a groove on one side of it a tough thread, with one end
fastened to the dish and the other to the body of the mus-

sel inside the shell. When the foot is withdrawn the thread remains and holds the mussel fast. Thread after thread is formed in this way until the mussel is fastened firmly enough to stand the hardest storms. When first hatched they swim about by means of cilia and later crawl by means of the foot, and even after they are firmly anchored are said to be able to let go and crawl about till they find a more suitable place. While the tide is out the mussels close their shells tightly and so keep themselves moist till it rises again, when they open a little and allow the water to pass in and out

Fig. 5.—*Modiola plicatula.*

through two fringed openings passing over the gills and carrying the food to the mouth. They open readily in confinement and the water may be seen passing in and out. The mussel does not live exclusively between tides, but grows larger farther down and often completely covers sand bars and shallow coves where there are stones enough for it to attach itself. It is also found attached to the roots of large seaweeds and specimens of a large size are often dredged in deep water. In this country the mussel is seldom eaten, perhaps on account of the abundance of oysters and clams which are generally preferred as articles of food; but in

Europe the same species is extensively eaten and is even cultivated for food. The American mussels are equally as good and if clams and oysters should ever become scarce would make a good substitute for them.

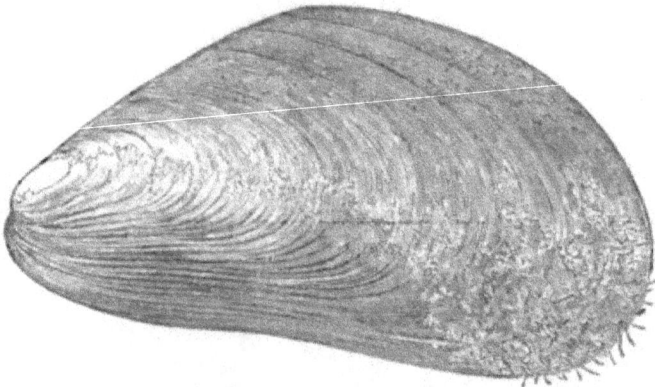

Fig. 6.—*Modiola modiolus.*

There is another mussel, *Modiola plicatula*, Fig. 5, with a fluted shell that lives along muddy shores nearly up to high water mark, and a larger species *Modiola modiolus*, Fig. 6,

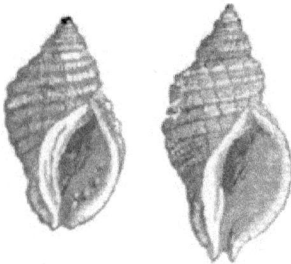

Fig. 7.—*Purpura lapillus.*

with a reddish body, which lives in deeper water and is usually found attached to the roots of the "devil's apron" thrown up by storms.

Several species of snails prefer to be out of water part of the time and if kept in an aquarium will crawl over the edge, and fall over on the outside where they will lie for several days without water apparently uninjured. One of the commonest of these is *Purpura lapillus*, a whitish shell, Fig. 7, that lives among the barnacles on which it feeds and there lays its eggs in oval capsules half an inch long, Fig. 8, fastened by one end to the stones. The shells of this snail are very variable in size and shape; some having a long spire and others a very short one, as shown in the figure. This is a northern species not found farther south than Long Island Sound. A larger gray shell, *Urosalpinx cinerea*, Fig. 9, lives in the same situations farther south, but north of Cape Cod is only occasionally found in warm bays or brackish rivers.

Fig. 8.—Egg cases of *Purpura lapillus*. Each case contains several eggs.

It lives on other mollusks and is very destructive to oysters, drilling through the shell and sucking out the

Fig. 9.—*Urosalpinx cinerea.*

contents. The apparatus by which this boring is effected is described farther on in the account of a larger snail of similar habits, *Lunatia heros*.

There are several snails of the genus Littorina which are very abundant between tides especially on the rockweed. The largest species, *Littorina litorea*, Fig. 10, has been introduced, or at any rate has spread southward, within comparatively few years. It is common in England, where it is known as the "periwinkle" and is eaten in large quantities. It was first noticed on the American coast at Halifax and has gradually spread southward and increased in abundance as far as Long Island Sound. It is not mentioned in the "Invertebrata of Vineyard Sound," pub-

Fig. 10.—Periwinkle. *Littorina litorea.*

lished in 1874, but has since been found at New Bedford, Mass., Watch Hill, R. I., and Stonington, Conn. The smaller species, *Littorina rudis* and *palliata*,

Figs. 11 and 12, are very common on the rockweed and their empty shells are mixed with sand up to high water mark, forming a large proportion of the water-worn shells which are gathered by patient summer visitors for the manu- facture of ghastly picture frames and other ornaments. All these species are worth eating, either raw or, better, slightly boiled.

Fig. 11.—*Litto- rina rudis.*   Fig. 12.—*Lit- torina pal- liata.*

On stones well down the shore, there is often found the "limpet," Fig. 13, a snail with a flat oval shell which it can draw down so close against the stones that it can hardly be pried off. The best way to see this snail to advantage is to keep it in a glass dish, through which the under side can be examined as it expands and creeps about.

Fig. 13.—Shell of Limpet, *Tectura testudinalis,* inside and profile.

Another species, almost as flat but with a short spiral at one end, is some- times found near low water attached to stones or to other shells. This is *Cre- pidula fornicata,* Fig. 14. It some- times remains so long in one place that the edge of its shell becomes irregular to fit the surface of the stone under it, but is said to leave its roost and wander about for food returning again to the same position. A still flatter species, *Crepidula plana,* Fig. 15, fastens itself often to other shells and is found inside those of dead snails with its own shell bent to fit the curve under it.

Fig. 13a.

There is another small flat shell, often found attached to
the under side of stones, that is white and shining as though
silvered. This is not the shell of a snail
but of a bivalve, *Anomia*, Fig. 16, like
an oyster attached by a stem on the
under side. The under valve is very
thin and flat and has near the hinge a
hole which surrounds the stem.

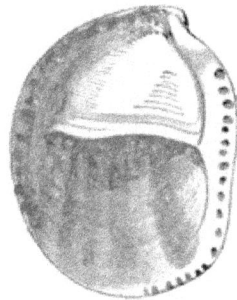

Many animals that live underground
prefer this region between tides. The
common clam is found here, buried deep

Fig. 14.—Shell of *Crepi-
dula fornicata*, under
side.

in sand or mud into which it burrows
by the help of the foot, the same organ
by which the mussel makes its threads.
The foot of the clam is flat, and is
pressed down into the mud and then expanded at
the end and
the shell

Fig. 15. — Shell of
Crepidula plana.

drawn down toward it. The
edges of the openings
through which the water
runs in and out are in the
clam extended into a long
double tube, commonly
called the head, which extends up to the surface of

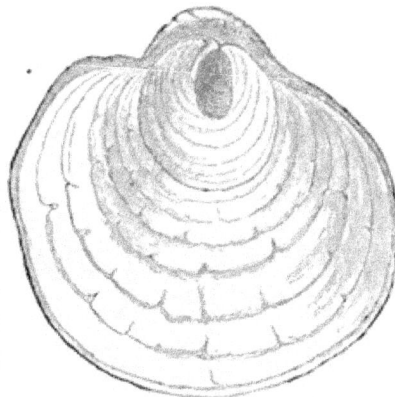

Fig. 16.—*Anomia*, under side, showing the
hole where the shell has grown round the
stem.

the sand and carries down water and food. Fig. 17 is a diagram of the inside of a clam, buried in the mud in its natural
position, with the large end of the shell downward and the

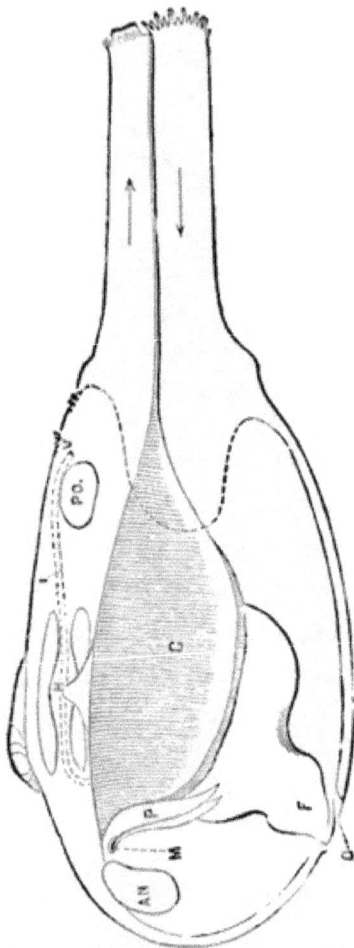

Fig. 17.—Diagram of the anatomy of the common clam. The arrows show the direction of the currents of water throu·h the tubes. *F*, the foot contracted; *G*, gills; *M*, mouth; *P*, palpi at the sides of the mouth; *A N*, *PO*, muscles which close the shell. [After Morse.]

tubes extended upward above the surface of the mud. The mouth of the clam is downward. The water runs in through the tube at the right, passes over the gills G, and by the mouth M, where the food of the clam is taken from it, and out again through the other tube. The razor clam, Fig. 18, lives buried in the same way in cleaner sand. It has a very strong foot and can burrow as fast as a man can dig. It can be found like the common clam by its hole, and must be dug by one quick stroke of the spade or it is sure to be lost.

There is a little clam, *Sax-icava*, Fig. 19, whose shell looks somewhat like that of the common clam, but it lives under and among stones and is almost always deformed by crowding between them; its tubes are colored red and are not united at the ends.

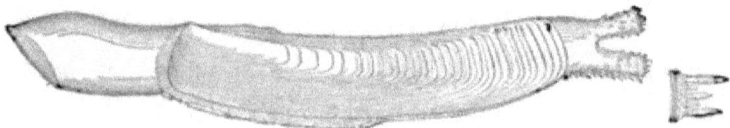

Fig. 18.—Razor Clam. [After Verrill.]

The common round clam, or quahaug, Fig. 20, lives farther down the shore where it is usually covered with water, and

Fig. 19. — *Saxicava arctica.*

is able to move about on the surface, or partly buried, by its large and strong foot. It is most abundant in warm and muddy bays, where it buries itself though not so deeply as the *Mya*, for its tubes, through which water passes in and out, are much shorter. This clam is used extensively for food, and south of Long Island Sound becomes the common clam of the markets replacing the more northern *Mya*. Farther north it is comparatively rare, but occurs in warm and sheltered places as far as the Gulf of St. Lawrence.

The "hen-clam," *Mactra solidissima*, Fig.

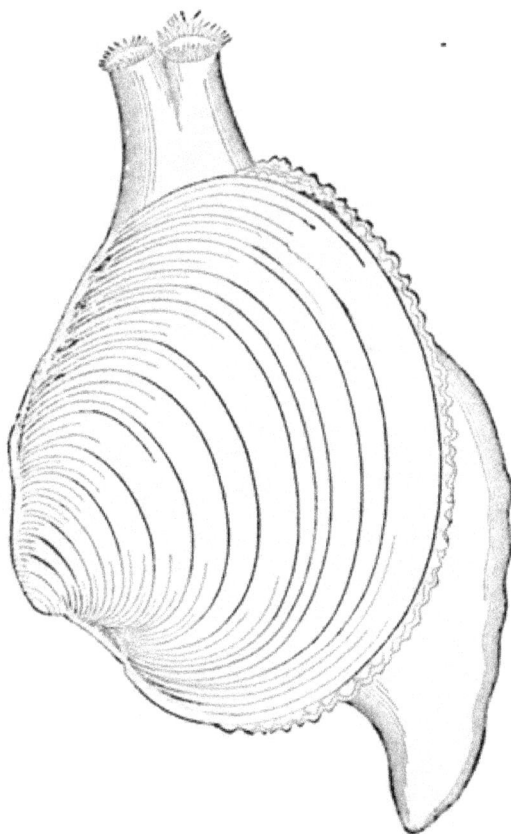

Fig. 20. — "Quahaug." *Venus mercenaria*, with the foot and tubes extended, and the edges of the mantle showing beyond the shell. [After Verrill.]

21, is a larger and more active species, living below low tide along sandy beaches and often thrown up in stormy weather, as it does not burrow very deeply, though it has a large and strong foot.

Fig. 21.—"Hen Clam," *Mactra solidissima*.

A delicate purple and white shell *Siliqua costata*, Fig. 22, is often found dead on sandy beaches, though its home is farther down below low tide.

On muddy shores, even in the dirty mud about docks

Fig. 22.— *Siliqua*.

and wharves lives a little round clam, *Macoma fusca*, Fig.

23. Its shell varies much in color and texture, being thin and white where the mud is sandy and clean, and rough and almost black on dirty flats.

On sandy and muddy shores there live also, buried, many species of worms. The common bait worm, *Nereis virens*, Fig. 24, is one of the most familiar ones and is a good example of its class. The body consists of a great number of segments, on each of which is a pair of complicated appendages, Fig. 25, consisting of thin paddles which serve for swimming and also as gills, and of bunches of bristles of various shapes.

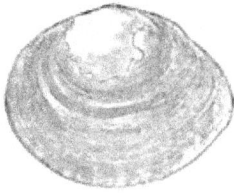

Fig. 23.—*Macoma fusca.*

At the head are several pairs of appendages of different kinds and two pairs of eyes. There are also long appendages on the tail. These worms have a pair of strong jaws which are usually drawn inside the mouth, but when in use they are protruded, as in Fig. 26, and can then give a strong bite. These worms live usually under ground in holes which are smoothed and hardened

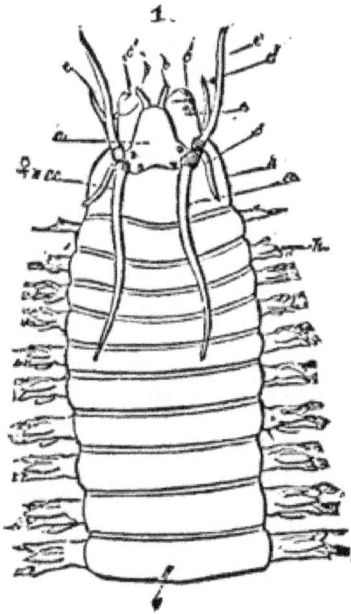

Fig. 24. — Head and front segments of *Nereis virens.*

by the slime from their bodies; but they come out at times, especially at night and in the breeding season, and swim about at the surface of the water.

Another very common beach worm, *Clymenella torquata*, Fig. 27, makes a tube of sand stuck together pretty strongly;

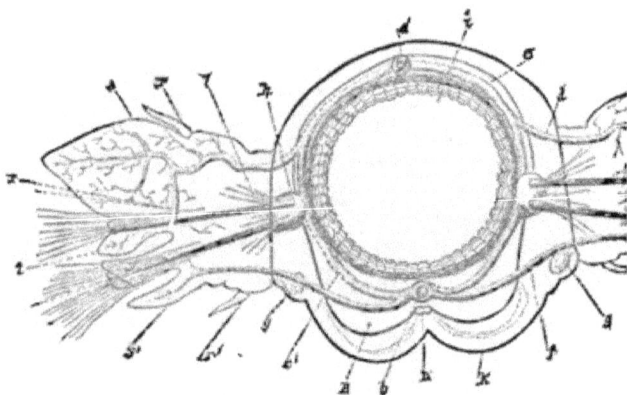

Fig. 25. — Section across one segment of *Nereis virens*, to show the appendages at the sides. It also shows the blood vessels running through them. [From Turnbull.]

and, if some mud is washed in a sieve till the finer part is washed through, great numbers of the tubes of this and

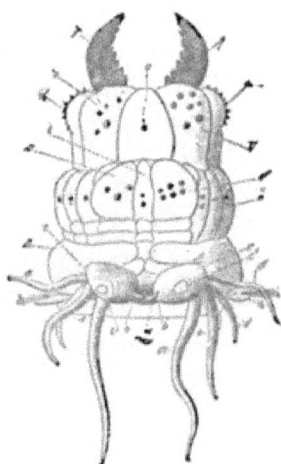

Fig. 26. — Head of *Nereis vi-rens*, with jaws extended. [From Turnbull.]

other worms will be found among the stones and shells that are left. Among these will often be found some very neatly made tubes larger at one end than the other. These are made by a brightly colored worm, *Cistenides Gouldii*, Fig. 28, which may sometimes be found in them, with only two large clusters of bristles visible in the larger end. A large part of the tubes dug at any time, however, prove to be empty.

In digging worms in mud, there is often found one that looks, when contracted, like a drop of blood; but if it is cleared of dirt and placed in clean water, it soon begins to extend its tentacles running them

Fig. 27. — *Clymenella torquata*, removed from its tube.

out like so many separate worms in all directions, Fig. 29.
With these tentacles it gathers in dirt to cover itself and

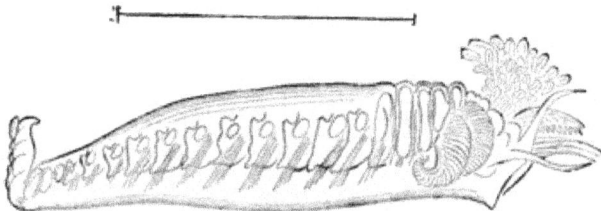

Fig. 28. — *Cistenides Gouldii*, removed from its tube.

will sweep the bottom of the dish clean.

Under stones on the beach and among roots of seaweed

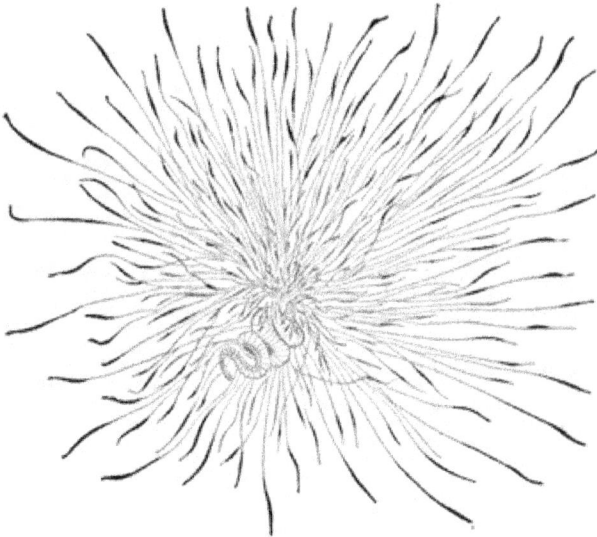

Fig. 29. — *Polycirrus eximius*, with its tentacles extended.

are generally found some short flat worms that resemble

centipedes at first sight. This resemblance is not so great if the under side is examined, for here the short segments and large number of paddles and bristles can be more distinctly seen. The back is disguised by flat scales attached to some of the segments and lying over one another, so as to cover the whole animal. The shape of the appendages on the sides and head can best be seen by removing some of these scales, as has been done in the worm, *Lepidonotus squamatus*, an illustration of

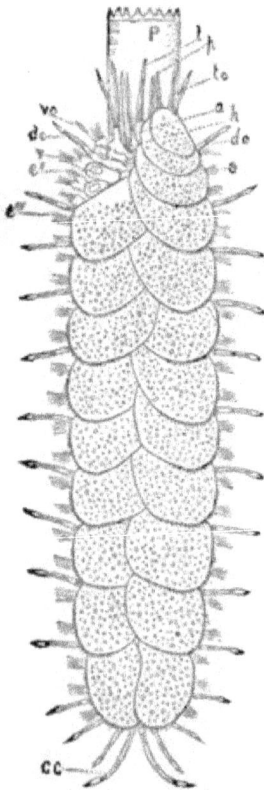

Fig. 30.—A Scaly Worm, *Lepidonatus squamatus*. Three scales have been removed from the left side to show the appendages near the head.

which is shown in Fig. 30.

Under the same stones will often be found worms of another kind, Nemertines, Fig. 31, without segments or side appendages.

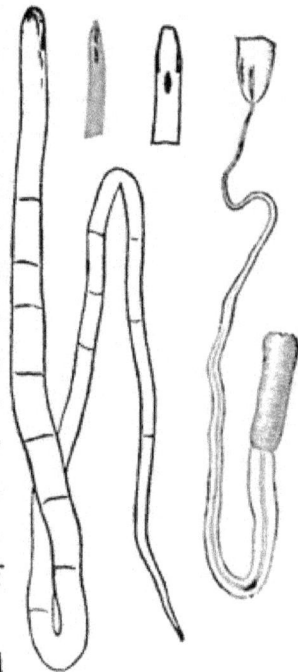

Fig. 31.—A Nemertine, *Lineus viridis*. The two small figures show the side and under surface of the head. The figure at the right is the proboscis extended.

They are olive-green or dull reddish-brown, and lie together in clusters stuck to the under sides of the stones. They are often twisted into the most irregular shapes and crawl by swelling out certain parts of the body and drawing toward them the parts behind. These worms have a peculiar organ, the proboscis, which is thrust out when they are injured, or from any cause contract suddenly.

Another Nemertine, *Cerebratulus ingens*, lives buried in the mud and is occasionally turned out in digging clams and worms. It is white and has been found as much as fifteen feet long and an inch broad. The proboscis is proportionally long and looks like another worm of a smaller kind.

Under the same stones occurs one of the flat worms, Planarians, half as broad as long, with two clusters of eyes at the forward end which are about all that distinguish one end from the other. These worms lie so close to the stones that they are apt to be overlooked, and they can be seen to a much better advantage when placed in a glass vessel where both sides of them are visible.

While these slow moving worms stick to the under sides of stones, great numbers of much livelier animals are often found on the mud beneath. These are small crustacea known as Amphipods. They are flattened sidewise so that they cannot stand upright but lie on their sides, and when disturbed kick about with feet and tail trying to get under something. In water they move much more gracefully, swimming rapidly by means of the paddles under their tails. As these are types of a large group of aquatic animals let us examine one more closely. Fig. 32 is *Gammarus ornatus*, one of the commonest species found under stones. Under the middle of the body are seven pairs of legs of three kinds : the two forward pairs having claws for grasping, the

next two pairs being short and the other three longer and
directed backward or turned up over the sides of the body.
Behind these middle legs are six other pairs, three of which
are thin and flat for swimming; while the three hinder ones
are short and stout and close together under the tail, with
which they form a powerful swimming or leaping organ. Under
the head are four more pairs of appendages used as jaws and
in front are two pairs of antennæ or feelers near the base of
which are the eyes.

Some of these Amphipods make tubes for themselves

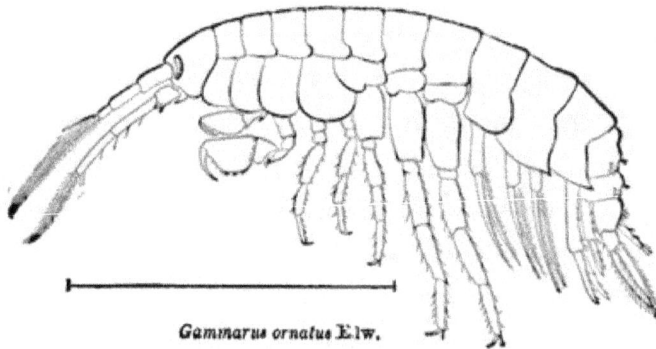

Gammarus ornatus Elw.

Fig. 32.

under stones; and, if taken out and placed in water, will
gather all the grains of sand and dirt they can find and
join them together until they get a pile large enough to hide
under. The little stones are joined together by threads,
like the threads of spiders and caterpillars, spun from the
two small pairs of legs under the middle of the body, which
are perforated at the end and contain glands which secrete
a fluid that hardens in water. Most of the Amphipods live
farther down the shore among seaweeds that are always
under water, and great numbers of them may be caught by

sweeping among these with a net; but there are a few
species, the sand fleas, that live up to or above high water
mark among the sand and rubbish. The common sand flea,
Orchestia agilis, is very abundant on sandy beaches, and
leaps about by the appendages near the tail, its gait resem-
bling that of the well known parasitic insect.

Small crustacea of another kind are often found under
stones along the shore, which are flattened downward instead
of sidewise. These are the Is-
opods, and resemble the garden
"sow bugs." They have the
same number of appendages as
the Amphipods, but the seven
pairs of feet of the middle of the
body are all alike. The swimming feet are closed up under
the tail and the front pair forms
a cover for the others.

Idotæa irrorata, Fig. 33, is a
common species, but is oftener
found among the eelgrass farther
out. It varies greatly in color,
some individuals being red, others

Idotæa irrorata Say.

Fig. 33.

dark green or yellow, and others again striped or spotted.
Idotæa phosphorea is oftener found under stones. It is more
pointed behind and is variously marked with gray and white
spots like the stones on which it lives. Another Isopod,
Sphæroma quadridentata is marked much in the same way
and lives under stones. It is short and oval and has the

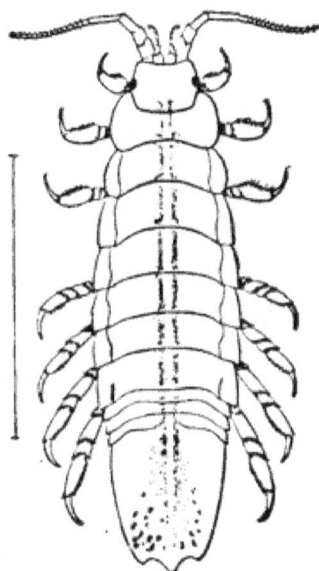

habit of rolling up in a ball, when threatened, and falling out of the way like some of the sow bugs.

When dredging once with a fisherman, he called our attention to the small flies which were gathered by hundreds under the sides of the boat just above the water ; and then, getting his broom, carefully swept them off because, as he said, they would turn into worms and eat into the wood of the boat. The flies in their young stages live swimming in the salt water, but the injury to the wood is caused by a small isopod, *Limnoria lignorum*, Fig. 34, which gnaws out minute holes as close together as they can be without breaking into each other. The wood thus wears and decays much faster than it otherwise would, the softer parts most rapidly, so that the knots are left projecting several inches beyond the surface of the timbers so eaten.

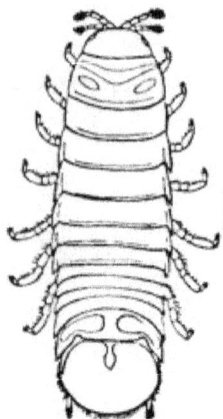

Fig. 34. — *Limnoria lignorum*, a wood-bearing Isopod.

Among the largest crustacea to be found between tides are the "fiddler-crabs," *Gelasimus*, Fig. 35. The males have one of the front claws very large and carry it across the front of the body, in somewhat the same position as the arm of a fiddler. The habits of one of these crabs, *Gelasimus pugilator*, are thus described by Mr. Smith.*

"They live on sandy beaches near high water mark, in holes in the sand, half an inch to an inch in diameter and

* Invertebrate Animals of Vineyard Sound, p. 42.

a foot or more in depth. Mr. Smith, by lying perfectly still
for some time in the sand, succeeded in witnessing their
mode of dig-
ging. In doing
this they drag
up pellets of
· moist sand,
which they car-
ry under the
three anterior
legs on the rear
side (the crab
walks side-
wise), climbing
out of their
burrows by
means of the
legs of the side
in front, aided
by the poste-
rior leg of the
other side. Af-
ter arriving at
the mouth of
their burrows
and taking a
cautious survey
of the land-

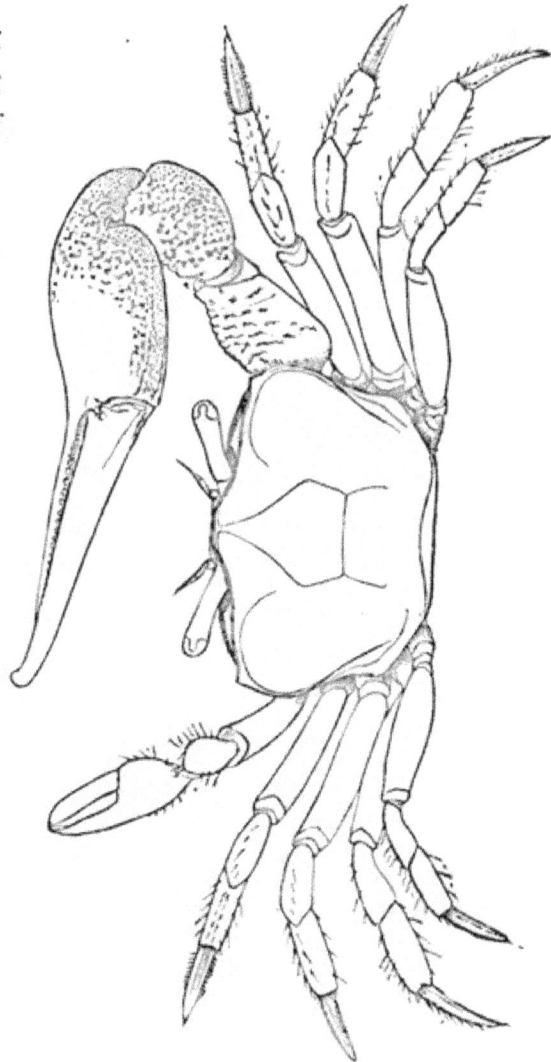

Fig. 35. — Fiddler Crab, *Gelasimus pugilator.*

scape, they run quickly to the distance often of four or five
feet from the burrow before dropping their load, using the

same legs as before and carrying the dirt in the same manner. They then take another careful survey of the surroundings, run nimbly back to the hole, and, after again turning their pedunculated eyes in every direction, suddenly disappear, soon to reappear with another load. They work in this way both in the night and in the brightest sunshine, whenever the tide is out and the weather is suitable. In coming out of or going into their burrows, either side may go in advance, but the male more commonly comes out with the large claw forward. According to Mr. Smith this species is a vegetarian, feeding upon the minute algæ which grow upon the moist sand. In feeding, the males use only the small claw with which they pick up bits of algæ daintily; the females use indifferently either of their small claws for this purpose. They always swallow more or less sand with their food. Mr. Smith also saw these crabs engaged in scraping up the surface of the sand, where covered with their favorite algæ, which they formed into pellets and carried into their holes, in the same way that they bring sand out,—doubtless storing it until needed for food, for he often found large quantities stored in the terminal chamber." These crabs are very hardy and easily kept in confinement in a small quantity of water in which they keep in incessant motion for months without taking any visible food.

ANIMALS BELOW LOW WATER MARK.

ANIMALS NEAR LOW WATER MARK.

FARTHER down the shore just below low water, or only uncovered at lowest tides, is a region which is inhabited by a greater variety of animals than that between tides, and will repay the trouble of wading and rowing along its borders. There live the crabs and lobsters, the latter well known at least in their dead and cooked condition, Fig. 36. If we compare a lobster with the small crustacea described in the last chapter, we shall notice that the joints of the middle portion of the body are covered up by a shell, which extends back over them from the head, and under the edges of this shell are a series of gills attached to the basal joints of the limbs. Five pairs of limbs are adapted for walking, the front pair of which have the large claws with which the lobster seizes its prey.

In front of the large claws are six pairs of smaller limbs used for chewing, and in front of these the eyes raised on short stalks and two pairs of antennæ or feelers, or, as fishermen call them, smellers. The hinder part of the body resembles more that of the other crustacea ; its joints are distinct and it has six pairs of limbs for swimming, the hinder pair of which are very large and broad, and form, with the last joint of the body, a powerful swimming or leaping organ which the lobster strikes forward suddenly when alarmed, and so darts backward into its hole. Young lobsters swim forward by the little feet under the tail, but the adults creep on the bottom most of the time. The whole body is covered with a shell so hard that the animal has to shed it from time to time as it grows larger. The shell cracks along the middle of the back, the front part separates from the tail and the limbs, even the large claws are drawn out in a soft and flabby condition. The skin of the eyes, the antennæ and the inside of the mouth and throat, are

Fig. 36.—Lobster, *Homarus Ameri-canus.*

all shed at these times. The limbs of lobsters and crabs break off easily, but are reproduced and after the next moult appear again, smaller than before but of the same shape and number of joints.

The eggs of lobsters are often to be seen on specimens in the market, being carried under the tail attached to the swimming feet until they hatch. The young lobsters can be easily seen through the egg, in the later stages, with the

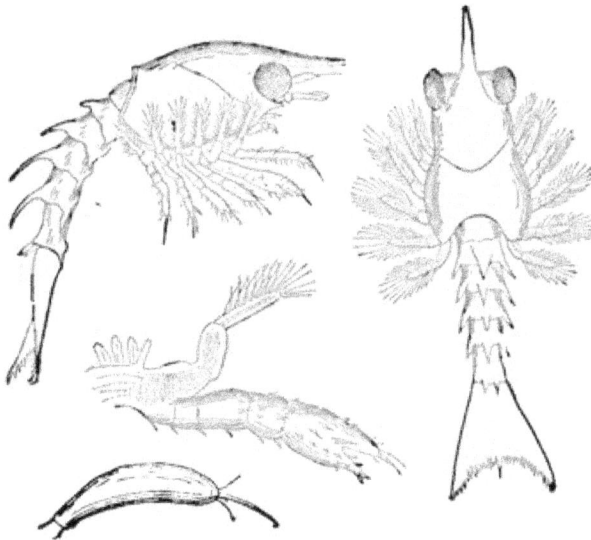

Fig. 37. — Young Lobsters; back and profile, below a leg and antenna.

feet packed in underneath them and the eyes much larger in proportion than in the adult. When they hatch they differ greatly from the adult both in shape and habits, Fig. 37. Their five pairs of limbs and the pair of foot jaws in front of them are all about of the same size, the big claws being only slightly larger than the others. All these limbs are two-branched, one of the branches turning outward and

upward above the other. The six pairs of feet under the
tail are also nearly equal in size, the hinder part not yet
enlarged as in the adult; while the terminal joint of the
body, which in the adult is narrowed toward the tip, is in
the young widened into a fish-tail shape.   On the back
of the abdomen, too, are five
spines on the middle of the
segments, that disappear altogether in the adult.   These
little lobsters swim at the
surface until they get the
shape and appearance of
the adult, and the larger part
are eaten by other animals
or killed in some way or
other.   When they get to be
an inch long they begin to
live more at the bottom, and
from this size upward are
often found under stones
even above low water mark.
As they grow up, however,
they keep farther down a-
mong the rocks that are
always well under water and
here the traps for them are

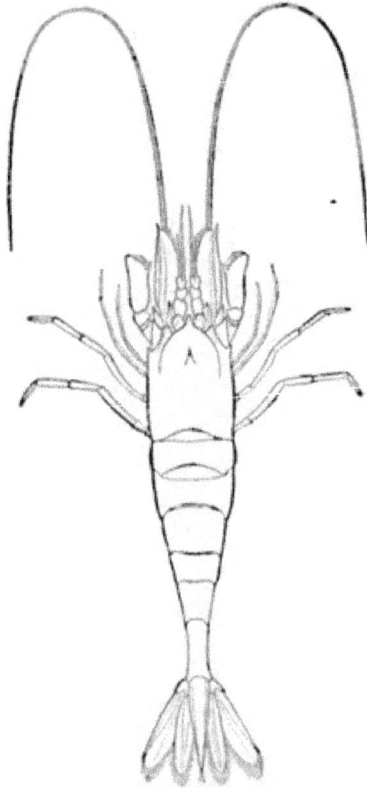

Fig. 38. — Common Shrimp, *Crangon vulgaris*. Natural size.

set.   These are cages made of laths with funnel-shaped
entrances like a rat trap, through which the lobsters go after
the bait of dead fish and have not sense enough to find
their way out again.

The common shrimp, *Crangon vulgaris*, Fig. 38, is often mistaken for the young of the lobster which it does not, however, much resemble. It lives on sandy or muddy shores and changes its color to imitate that of the bottom on which

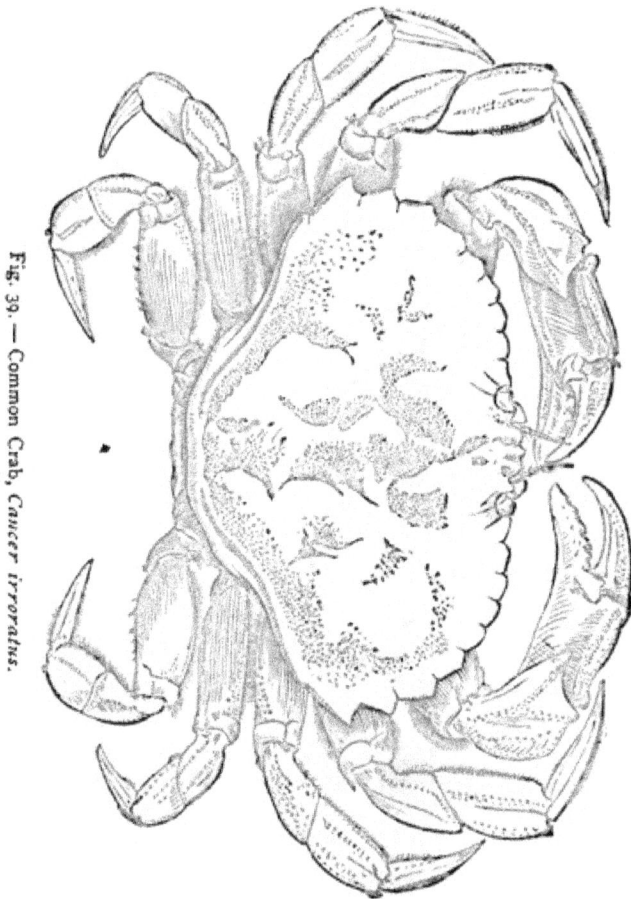

Fig. 39. — Common Crab, *Cancer irroratus*.

it rests. A dark individual, if placed in a white bowl, will soon become much lighter. This shrimp does not grow much larger than the figure and much smaller ones are often found with eggs.

The crabs may be considered as shortened lobsters. The middle portion of their bodies is covered up by a similar shell under the edges of which are the basal joints of the legs with the gills attached. The tail is very small and folded under the middle portion of the body, but if it is turned outward the same appendages are found as in the

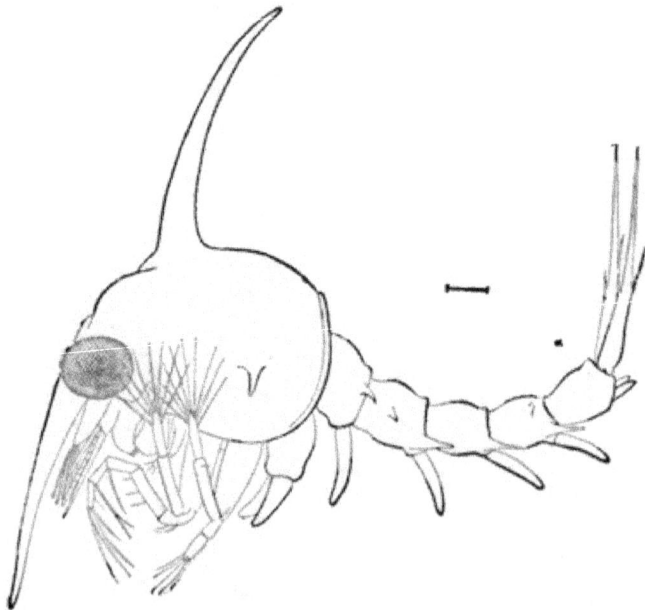

Fig. 40. — Young Crab, " *Zoea.*"

lobster. The same number of jaws and palpi are also attached to the head. The crabs are usually wider than long and the habit of walking sidewise is common. The best known crab along the whole coast is *Cancer irroratus*, Fig. 39. It lives not far from the surface of the water, occasionally above low water mark under stones and seaweed, and sometimes completely buried in mud. In its growth from the egg this crab goes through some very curious changes, differing so much, in its early stages, from th: a lult that the young were long considered as belonging to entirely distinct species. The eggs are carried under the tail like those of lobsters until they hatch and become little swimming animals like Fig. 40, but not more than a tenth of an inch in length. Its legs have not yet grown, and it swims in a jerky way by the front appendages which are eventually to become mouth organs. As it becomes older, the legs appear just behind these swimming appendages, between them and the tail, and become gradually larger until the next metamorpho-

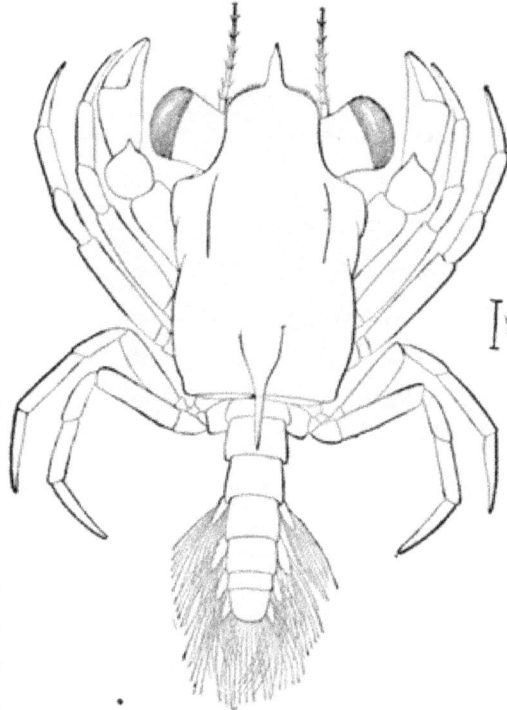

Fig. 41. — Young Crab, "*Megalops*.

sis. The tail is as large in proportion as that of a young lobster and carried extended behind. These little crabs swim at the surface of the water, and if kept in a dish collect around the edges keeping in constant motion, crowding each other out of the way with the sharp points on their heads. They are nearly transparent and sometimes can only be found by their dark eyes, and spots on various parts of the body. These little crabs moult several times in this stage, but finally a great change takes place and they become like Fig. 41, a little more like the adult crab. They now have their five pairs of legs; but the eyes are still enormously large, there are long spines on the back, the abdomen is still carried extended, and they have swimming feet which are used like those of the lobster. They still swim at the surface, but rest on the bottom at times and are soon ready for another metamorphosis. In the next stage, the habit of swimming is given up, the tail is turned under the body and the little crab lives on the bottom like an adult; but still the body is longer and the shape very different from the old crabs and it is only after several more moults that the final form is reached. A similar series of change is gone through by the other crabs, though they have not been so well made out as in this species. The nearest relative of this crab is *Cancer borealis*, which is found with and sometimes mistaken for it. It is about the same size but is more oval, thicker, and much more peaceful in its habits. When the tides are unusually low, some of these crabs are occasionally found holding on the rocks in the most exposed places, stupidly waiting for the water to rise and cover them instead of following it down or hiding comfortably under damp seaweed as the other species does. They will allow themselves to be picked up without showing fight and will lie quietly for hours, even if several are piled up together, in this respect differing decidedly from the common kind.

The mud crabs, *Panopeus*, Fig. 42, are found among stones or buried in mud, and when caught stretch out their legs stiffly as far as they can, so as to make themselves as unpleasant to handle as possible. They are of small size and are found along the coast as far north as Cape Cod.

The "oyster crab" is carried wherever oysters are. The female lives inside the shells of oysters and is often found among oysters sold for food and cooked with them, but the males are rare and occasionally found swimming free in the water.

The "blue crab," Fig. 43, which is much eaten in New York and southward, and occasionally found north of Cape Cod, lives also near low water on muddy shores and among eelgrass where it often conceals itself in the mud. Unlike the common shore crabs, *Cancer*, it swims readily by the help of paddles on its hind legs. For eating, the blue crabs are preferred "soft-shelled,"

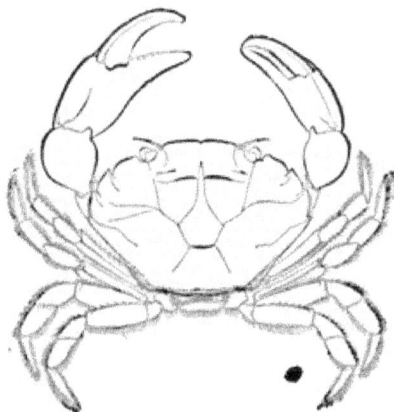

Fig. 42. — *Panopeus depressus.*
Mud crab.

that is, after they have moulted and before the new skin has hardened. They seem to moult at various times during the summer and to remain soft a considerable time.

There is another swimming crab, *Platyonichus ocellatus,* Fig. 44, found from Cape Cod southward very different from the last. It has a rounded body and is white with the back covered with red and purple rings. It swims rapidly at the surface, even when full grown, but lives usually near the sandy bottom, in which it can quickly bury itself for con-

cealment, or to escape being washed ashore in rough weather.
It feeds on dead animals of any kind and soon finds anything decaying that may be left in the water.

The "spider crab," Fig. 45, is a round long-leggèd species

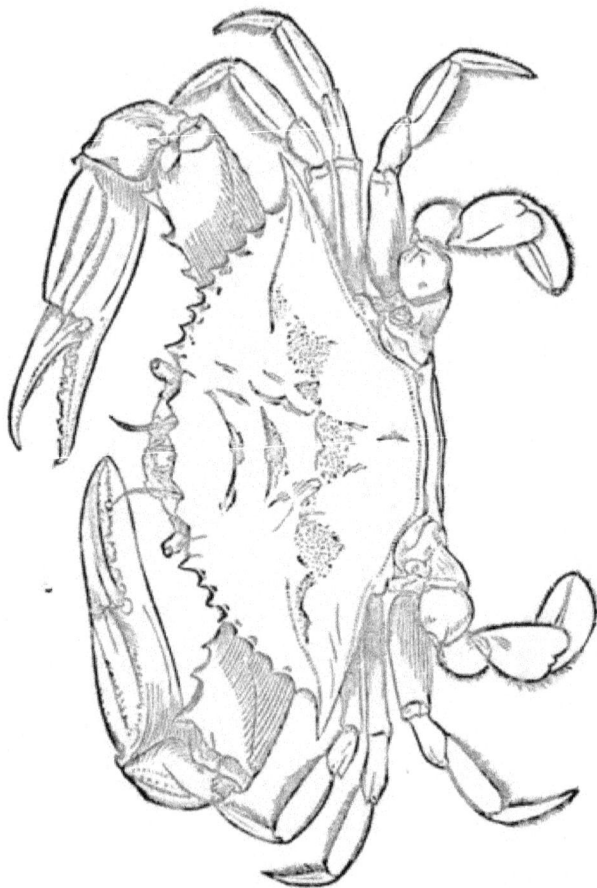

Fig. 43.—Blue crab, *Callinectes hastatus*.

that lives among mud and weeds and has a coat of hair that
entangles dirt of all kinds, furnishing a place of attachment
for seaweeds and hydroids which help still more to conceal
the animal. It grows to be a foot or more across.

The "hermit crabs," Fig. 46, are among the most amusing animals of this group, on account of their peculiar habit of covering the hinder part of their bodies with a snail shell, or any other hollow object which they can carry about with them. Like the crabs, they swim at the surface when young, but at an early age settle to the bottom and begin to live in

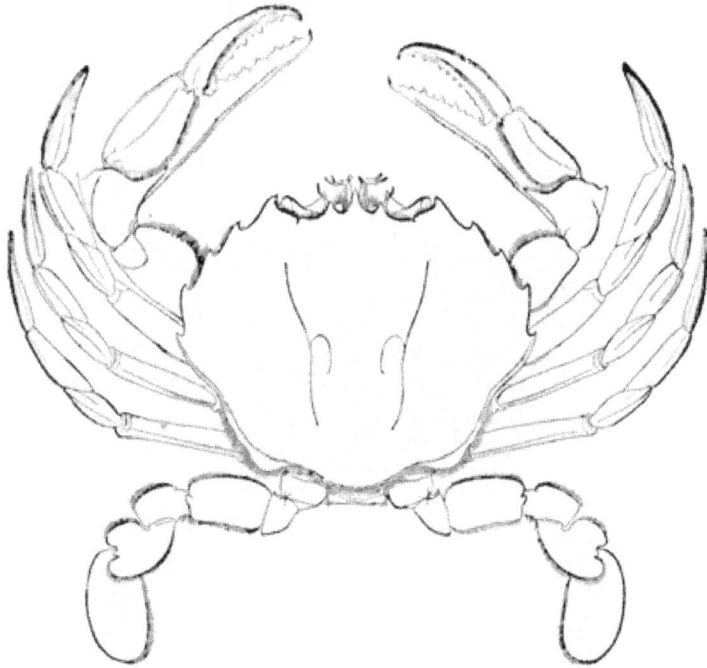

Fig. 44. — Lady crab, *Platyonichus ocellatus.*

deserted shells. Their tails are soft and have strong hooked appendages at the end for holding into the shell. Their legs are crowded forward, so as to project from the shell and enable the crabs to walk about, dragging the shell after them, without exposing much of their bodies. As they increase in size they have to leave their snail shells and find larger ones, going about from shell to shell until they are fitted. These crabs are hard to keep in confinement, as they need a good deal of clean water to keep them healthy.

The Amphipods and Isopods mentioned in the last chapter are still more abundant down among the seaweeds, and among them occurs a curious form, *Caprella*, Fig. 47, which at first sight would not be recognized as related to them. These *Caprellas* are very common on the eelgrass, holding

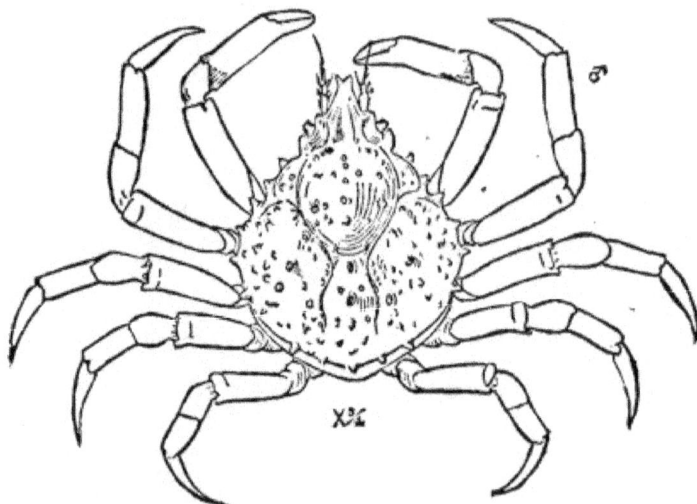

Fig. 45.—Spider crab, *Libinia canaliculata*.

on by the hinder legs like canker worms and swinging the forward part of the body up and down.   The middle legs are rudimentary.

The horseshoe crab, Fig. 48, is common along muddy and sandy shores, and in the early summer pairs of them come up on the shores to lay their eggs in the sand. The cast-off skins of these animals are often found along the shores of muddy rivers. The horseshoe crab is interesting as being the last surviving relative of a group of animals which have died out and are now only known in a fossil state. It differs greatly from the crabs and resembles in some respects the spiders, like which it has six pairs of limbs and no antennæ.

There are other curious animals often found under stones, and among seaweeds and hydroids, that are nearly all legs, Fig. 49, known as *Pycnogonidæ*. Most of them live on hydroids. ·The eggs are carried by the mother, attached in balls around the front limbs until they hatch; then the young, which differ considerably from the adult, attach themselves to the hydroids and go through metamorphoses very much like those of some parasitic mites, like the red mite found on grasshoppers, for instance. The body of the adult is so shaped that there is no room for the in-

Fig. 46. — Hermit Crab, *Eupagurus bernhardus*, in a shell of *Lunatia heros*.

ternal organs, except in the legs and branches of the stomach, and the eggs extend into these.

On sandy mud, just below low water, lives one of the largest snails of the coast, *Lunatia heros*, Fig. 50. When in motion, the soft part of the body is extended out of the shell to three or four times the size of the latter, Fig. 51, so that the shell is nearly covered up; but if the animal is touched it squeezes out the water and in a few moments draws into the shell again, covering itself with the horny operculum which it carries on its back. It creeps about partly buried in the sand searching for its food. It lives on other snails and bivalve mollusks, through whose

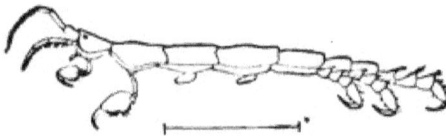

Fig. 47.—*Caprella geometrica*, enlarged.

shell it drills by means of the teeth on the tongue, Fig. 52, and scrapes out the contents through the hole. The eggs of this snail, which are laid in a ring mixed with sand, Fig. 53, may often be picked up on the beach. If one of these be held up to the light the egg cases can be seen as transparent round spots in each of which are several eggs.

On sandy shores great numbers of the little snail, *Tritia trivittata*, Fig. 54, are often found creeping about through the sand leaving shallow furrows after them. They are very active animals for snails, and when kept in clean water will sometimes stretch out to their full length and jump about the dish. On more muddy flats another species, *Ilyanassa obseleta*, Fig. 55, is generally to be found following the

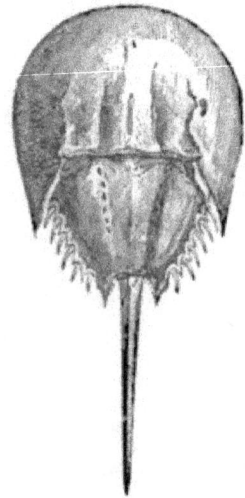

Fig. 48.—*Limulus polyphemus.* [From Tenney's Zoology.]

same habits. Both species may be seen covering dead fish or anything decaying and are of great use as scavengers.

A little brown shelled snail, *Locuna vincta*, Fig. 56, is very common on the eelgrass and makes the ring-shaped clusters of eggs so common there.

The branching plant-like growths along the shore are most of them not plants at all but hydroids, stationary animals related to the jelly-fishes and some of them being the young of jelly-fishes (an account of which will be given further on). Among these hydroids is the favorite hiding place of many

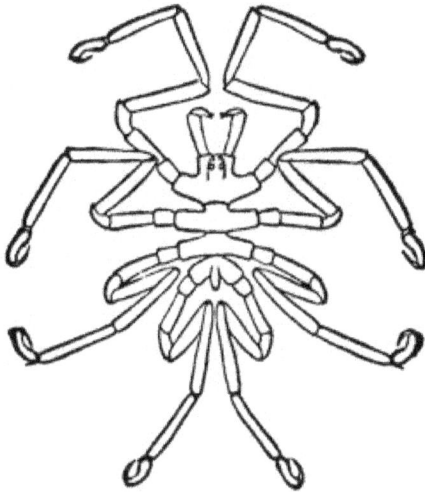

Fig. 49.

of the "Nudibranchs," or naked gilled snails. These are often brightly colored and have no shells ; but, in place of them, appendages of various shapes along the back, which make them resemble the branching objects among which they live. Their eggs are laid in lumps of transparent jelly attached to plants and hydroids and the young hatch as many snails do, with little shells with one or two spirals and with two lobes on the sides of the head, on the edges of which are cilia by which the animals swim, Fig. 57. As they get older they lose the ciliated lobes and begin to creep on the foot like old snails; but the nudibranchs, at this time,

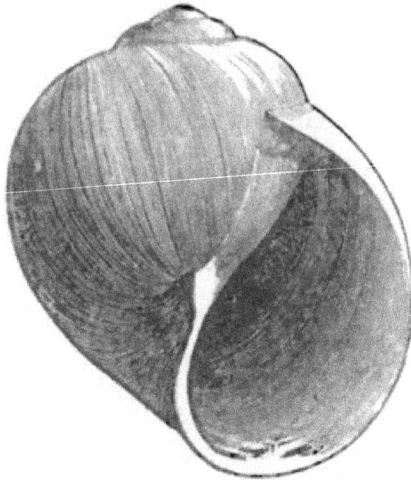

Fig. 50.—Shell of *Lunatia heros.*

lose also the shell and have on the back the various appendages which have been mentioned. The largest native species is *Dendronotus arborescens*, Fig. 58, a long brownish snail with tree-shaped appendages along the back that make it look like a mass of seaweed. This resemblance is still greater when the animal shortens itself and gathers the branches together.

Several species of Æolis, Fig. 59, are transparent white

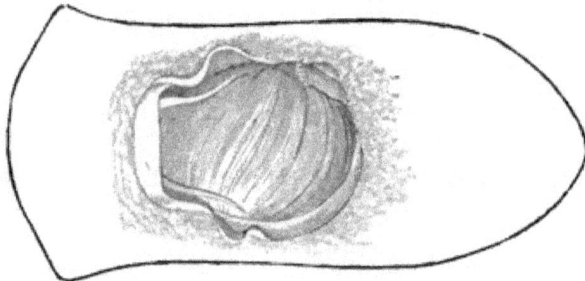

Fig. 51.—*Lunatia heros* crawling, fully expanded.

with red or brownish papillæ arranged in rows on each side of the body. Another species, *Doto coronata* has a few club-shaped red appendages covered with black spots. The genus Doris has a rosette of soft appendages at the posterior end of the body, Fig. 60. The common *Doris pallida* is white and has the back covered with white knobs. In cool weather these naked snails can be kept easily in confinement for several days. They are very hard to preserve in any way, and, when contracted by alcohol, many different species are so much alike that they can only be distinguished by the teeth on their tongues, which are the only hard parts to their bodies.

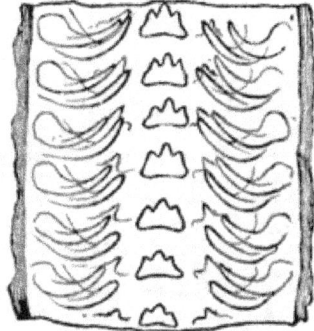

Fig. 52.

Several shallow water bivalves have been mentioned in connection with the clam in the last chapter, but none is better known than the oyster. This mollusk grows naturally from Cape Cod southward and is found in places farther

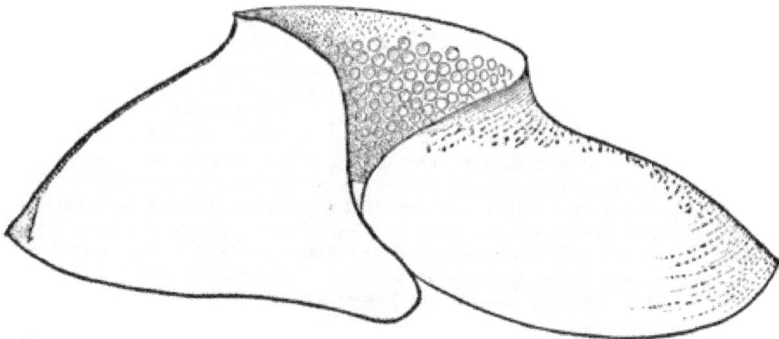

Fig. 53. — Ring of sand containing eggs of *Lunatia heros.*

north as far as the gulf of St. Lawrence, and old shells are found at so many places along the coast that it is supposed to have lived formerly at many more points north of Cape

4

Cod. The oysters, now brought from southern waters and kept in beds north of this point, to fatten and be taken up as wanted, do not breed there on account of the coldness of the water or some other cause. The oyster eggs are very small and are discharged into the water where they develop in a few days into little swimming larvæ, Fig. 61. Rudiments of the shell soon appear and the young, as further developed, is shown in Figs. 62 and 63. They soon begin to settle to the bottom and those which fall in suitable places become attached by the edge of the shell which, as it grows, cements itself to stones or shells, or any solid substance on the bottom, Fig. 64. The young which settle on the mud come to nothing, and those which attach in shallow water which is frozen in winter are all killed off before the next season, so that not one egg in a million produces a full grown oyster. Although the young become attached and start best on a hard bottom, the half grown oysters fatten better on muddy bottoms, full of microscopic plants, and it is therefore customary to dredge up young oysters and "plant" them on muddy grounds where they are left to grow a year or two before being gathered for the market.

The differences between various kinds of oysters depend much on the kind of bottom on which they fatten; the best places being generally shallow bays and mouths of rivers where the water is warm and brackish. The shells of oysters furnish good attachments

Fig. 54.— *Tritia tri. vittata.*

Fig. 55.— *Ilyanassa obsoleta*

Fig. 56.— *Lacuna vincta.*

for their own young, and for many other animals which are carried with the oysters to distant parts of the coast where they do not naturally grow.

The shells of oysters, like those of other mollusks, are formed from the mantle, as it is called, a fold of skin that lines the shell on the inside. This secretes from its whole surface the pearly lining of the shell, and from its outer edge the harder part that is added to the outside which usually shows lines of growth, like the rings in a tree, where the increase has not taken place at a uniform rate. Oyster shells also show their age by the deeper grooves between the portions of the shell, formed in different seasons which in some shells are very distinct. Pearls are formed when sand or any foreign substance gets

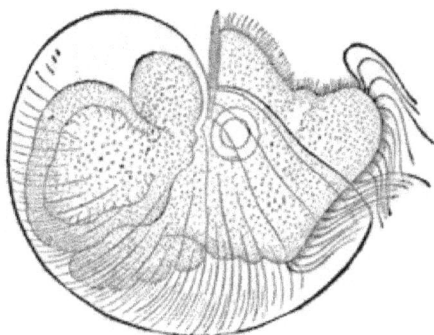

Fig. 57.—Young of *Eolis diversa*.

between the mantle and the shell, causing the secretion of the shell lining to go on more rapidly at that point, and the beauty of the pearl depends on the kind of lining which the shell naturally has, those of the oyster being dull and opaque like the rest of the shell.

The common "scollop," *Pecten irradians*, Fig. 65, lives along the shore as far north as Cape Cod, on muddy bottoms and among the eelgrass, where it lies at rest with the shell slightly open, showing the thickened edges of the mantle, fringed above and below with tentacles among the bases of which are two rows of eyes extending round the shell. The scollops can swim by opening the shell and closing it suddenly, driving themselves backward. The muscle which closes the shell is large and strong and is the part which is used for food, the rest of the scollop being thrown away. There is a much larger species, *Pecten tenuicostatus*, with a nearly smooth shell that lives farther north and is often dredged in deep water in Massachusetts bay and on the coast of Maine.

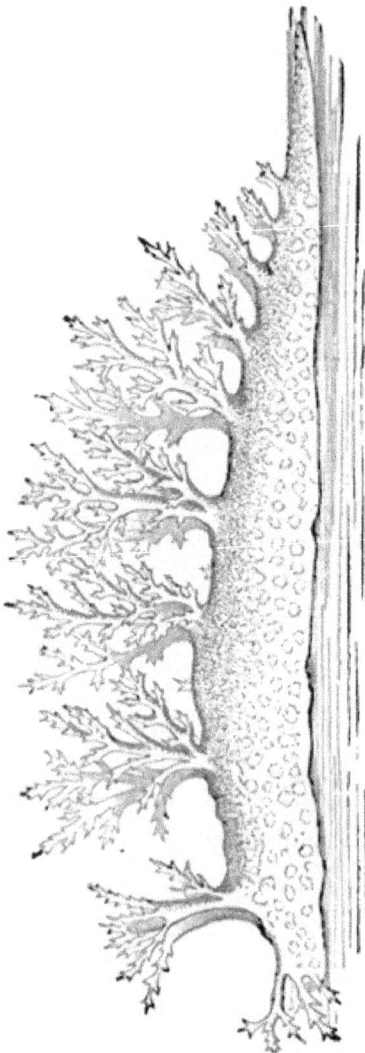

There is a very curious bivalve, *Teredo*, Fig. 66, that burrows into wood just below low water and does great damage to ships, buoys and all kinds of wood-work under water. The holes are small at the surface of the wood, where they are begun by the animals when young, but as they grow they dig deeper and enlarge these holes to a quarter of an inch in diameter. The holes run through the wood in all directions but

Fig. 58.—Dendronotus arborescens.

never interfere with each other, twisting about wherever they find room. The eggs remain in the gill cavities of their parents until they develop into ciliated larvæ, which can swim about in the water, and soon have a bivalve shell large enough to cover them. They then begin to attach themselves and, as soon as they find suitable wood, to bore into it. The shells always remain very small and the animal grows chiefly in length, so that it is popularly considered a worm. The tubes through which water runs in and out are small and must, of course, always be near the outer opening of the burrow. The Te-redo does not eat the wood which it digs out,

Fig. 59. — *Eolis pilata.*

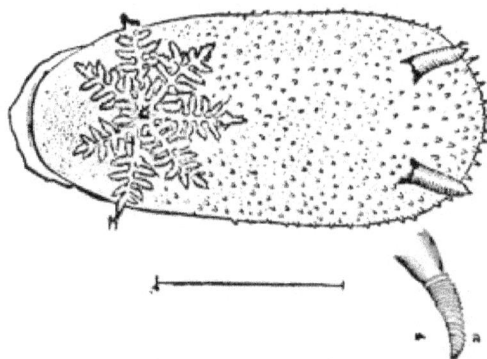

but gets its food from the water which comes in through its tubes, just as the clam does; so that paints, or other poisonous substances in the wood, are not much protection

Fig. 60. — *Doris bifida.*

against it, and it can only be stopped by plates of copper or some such substance too hard for them to bore through. The wooden buoys used to mark the channels into harbors

are painted with copper paint, but are attacked wherever
the paint is rubbed off and in course of time destroyed by
these animals.

The worms, which have been mentioned as occurring on
beaches, and many others, live also farther down below low
tide and may be found by digging or dredging ; but there
are a few which live above ground and can hardly fail to be
seen while looking for other animals. The eelgrass, rockweed and other water-plants, are often covered with little

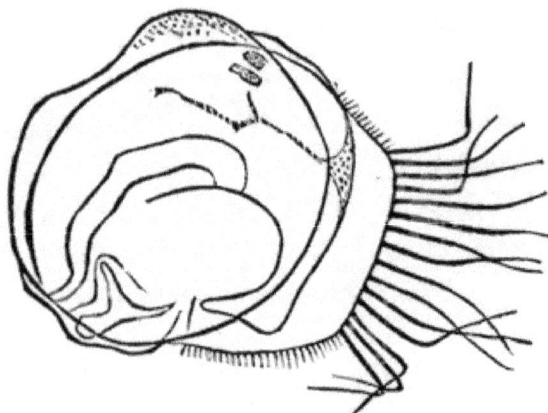

Fig. 61.—Young of Oyster. [From Brooks.]

white spiral shells that are easily mistaken for those of snails
while they are really made by worms, *Spirorbis*. If some
of these are placed in a dish of water the heads and gills
of the worms will soon be seen to protrude. The head is
surrounded by a collar from which extend on each side
feather-like gills often brightly colored. One of the branches,
however, is club-shaped, and when the worm contracts is
drawn in last, closing the mouth of the shell. On the eelgrass there is often found a small yellow worm, *Nicolea sim-*

plex, which lives in thin tubes covered with dirt with only the long appendages of the head extended. It sometimes comes out of the tube and can swim about at the surface where it sometimes gets into the net with swimming animals.

In tubes among mussels and hydroids there lives a little worm, half an inch long, *Fabricia Leidyi*, which often creeps out when these are kept a short time in standing water. It has six feather-like gills on the head and at the base of them

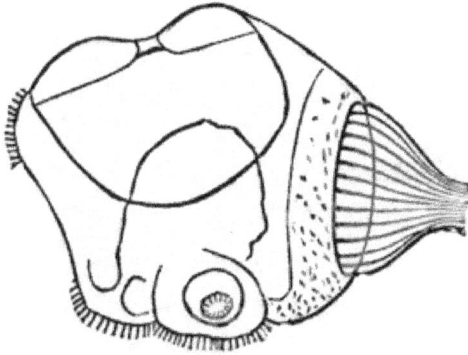

Fig. 62.—Young Oyster. [From Brooks.]

a pair of eyes, and it also has a pair of eyes on the opposite end of the body.

In the same places lives another worm which will be mentioned again, *Autolytus*. It makes thin tubes covered with dirt but, in confinement, soon comes out of them and spreads round the dish. It has a large number of segments and long appendages to each, and two pairs of eyes and several long tentacles at the head. One curious thing about this worm is that it is often found dividing into two worms. One of the segments near the middle gradually takes the form of a head with eyes and tentacles and in course of time separates from the part of the worm in front. These tube-living

worms do not produce eggs; but those which drop off from
them behind develop into perfect males and females and
swim about in the water (as will be noticed more fully hereafter).

Among stones and mussels are often found lumps of gelat-

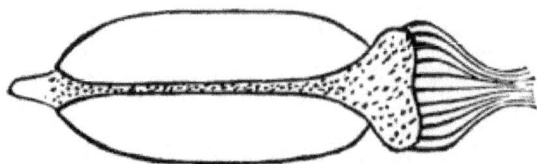

Fig. 63.—Young Oyster, seen edgewise. [From Brooks.]

inous looking animals, covered with dirt and showing no
signs of motion except by two holes in each, which can be
opened and closed and from which streams of water can
be ejected. They need a good deal of washing to clean
them so that their real shape
can be seen, but when cleaned
off, they look like Fig. 67,
Molgula manhattensis, which
is one of the commonest of
these ascidians. When young
they have a tadpole shape,
Fig. 68, and swim in the
water; but become attached
quite early, lose the tail and
change their form entirely.
They have two openings to

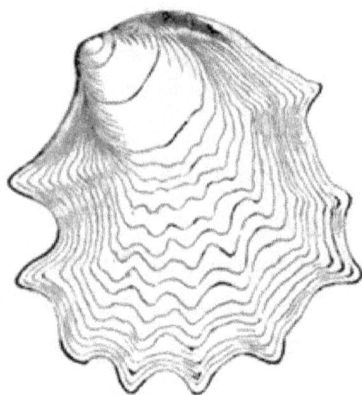

Fig. 64.—Young Oyster. [After
Verrill.]

the body, one the mouth into which the water passes into
the throat and through openings in its sides into a large

cavity, from which it escapes by the other external opening. The food passes down the throat into the stomach.

Other species live singly, attached to stones : as *Cynthia carnea*, which is bright red and has two brighter spots around the openings.   *Cynthia echinata* is round and is covered with branching spines which gather dirt enough to cover it completely.   There are other ascidians which grow in com-

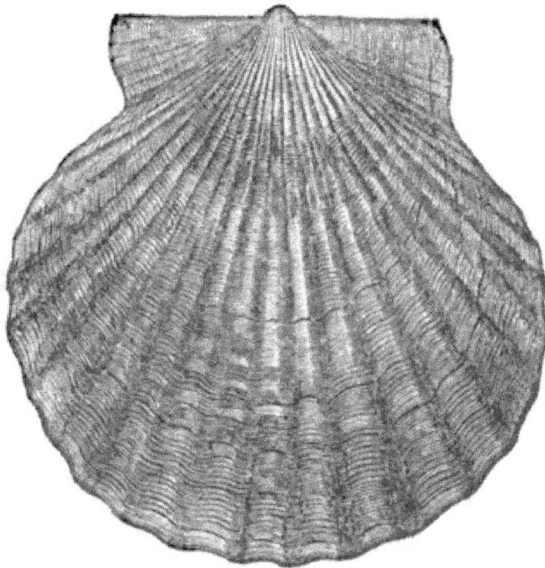

Fig. 65. — Scollop, *Pecten irradians.*

pound clusters and several of these are found near low water. One, *Batryllus Gouldii*, is in circular clusters, in the centre of each of which is a common discharge opening for the water from the respiratory cavities of all the members.   At first there is only a single circle, but they increase rapidly by branching and at length form large clusters of hundreds of circles, all forming one jelly-like mass often covering up entirely the leaves of eelgrass, on which it grows, with a brown or greenish crust quarter of an inch thick.

Little white clusters are often found on eelgrass and on stones, which are colonies of "polyzoa," Figs. 69, 71 and 72. These start from single individuals that multiply by branching in all directions, forming groups of hundreds, all more or less connected together. These things need a microscope and a good light to show them to advantage. They can be examined alive if allowed to stand quietly in

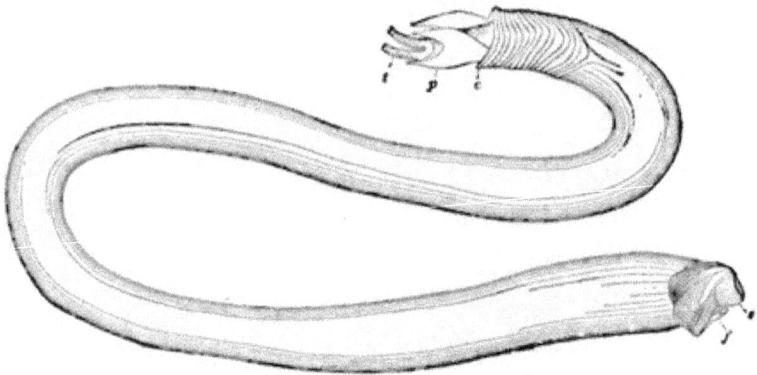

Fig. 66. — Ship worm, *Teredo navalis.* [After Verrill.]

water, when they will put out the ends of their bodies with the mouth and circle of tentacles, Fig. 70. The shells have usually a large opening, through which the animal expands, and several small openings. They vary, however, in different species, Fig. 73. Some kinds grow in branching clusters instead of flat, as *Bugula turrita,* Fig. 72, in a spiral. Others branch in a fan-shape. On these branching kinds it is easier to see the little objects, like birds' heads, that are attached to each shell and open and shut without any apparent object, Fig. 74. Other species have movable spines that swing up and down in the same way. Some polyzoa do not make hard shells like those which form the white clusters on stones, but soft and flexible ones. A very common one makes brown clusters on the rockweed and especially about its roots. The young polyzoa just from the egg have no resemblance to the adult and swim about by cilia, Figs. 75, 76 and 77. As they grow larger they have a bi-valve shell which remains till they have become attached and begun to branch into a compound cluster.

Fig. 67.—A simple Ascidian, *Molgula*.

The triangular larva, Fig. 75, is the young of *Membranipora pilosa*, Fig. 69 ; a very common species, Fig. 76, is the same seen edgewise. The other larva, Fig. 77, is of *Flustrella hispida*, one of the soft-shelled polyzoa that form clusters around rockweed. Both figures are copied from ·Barrois' Embryology of the Polyzoa.

The starfishes are among the most peculiar animals of the

Fig. 68.—A young Ascidian.

seashore and belong to a class, the Echinoderms, more of which live on land or in fresh water. The common star-fishes, *Asterias forbesii*, Fig. 78, and *Asterias vulgaris*, live near low water mark, coming above it occasionally and in winter retreating to deeper water. They live on mollusks

and are a great nuisance to the oyster growers. They fold
themselves around an oyster or mussel, turn their stomach

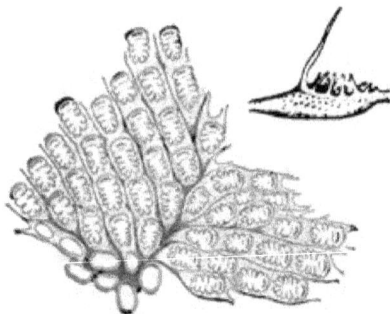

Fig. 69.—Polyzoa, *Membranipora pilosa*,
much enlarged. At the left upper corner
is a profile view of one cell.

Fig. 71.—Polyzoa, *Crisia eburnea*.
Round cluster enlarged.

out of the mouth and in between the shells of the bivalve
and digest it without taking it inside their bodies. The
starfishes move
by suckers in
the fine grooves
on the under
side of their
arms. To bring
them into use
they have to be
filled with
water from the
water tubes,
which receive
their supply
from the po-
rous colored

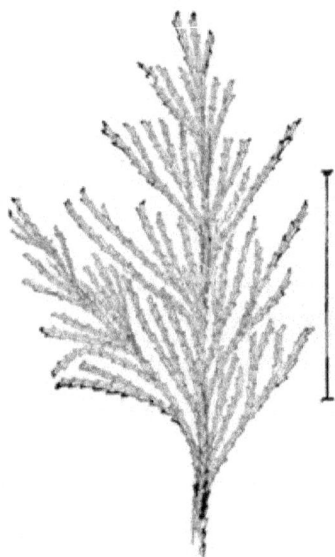

Fig. 72.—Polyzoa, *Bugula turrita*.
Enlarged twice.

spot on the
back of the star-
fish and carry it
through all the
arms giving off
a branch to
each sucker.

The skin of
the starfish is
filled with little
hard plates and
from it project
spines of vari-
ous shapes
which have,

around the base, clusters of little organs called pedicellariæ,

Fig. 79, which, like the similar organs on polyzoa, have jaws that open and shut for no apparent purpose unless to prevent dirt from sticking to the skin. At the end of each arm is an eye.

The development of the starfish is very complicated; the eggs are laid loose in the water, and grow into little larva-like worms that move about in the water by means of cilia

Fig. 70. — Polyzoa. A single animal of *Membranipora pilosa* expanded, much enlarged.

and have no trace of any radiate structure about them, Fig. 80. As they grow larger, they become more complicated in shape having long arms with lines of cilia running along them, Fig. 81. The internal arrangement becomes more complicated at the same time which may be easily seen as the whole animal is transparent. While it is very young, a pair of tubes begin to form at the sides of the stomach, which grow larger and larger becoming separate from the stomach and opening outward by a hole in the middle of the back. The first traces of the arms of the starfish appear on these water tubes, each side of the stomach, as five little lobes in a row along the tube, Fig. 81. As these grow larger the rest of the larva at length begins to grow smaller and finally dis-

Fig. 73. — Polyzoa, One branch of *Crisia eburnea*, much enlarged.

appears altogether inside the five lobes, which, instead of remaining in a row have bent around so that the end ones unite and form a star. The hole which lets water into the circulating tubes in the middle of the back of the larva becomes the colored porous spot between the two arms which close together last.

Besides the common starfishes there is a smaller smooth species, *Cribrella*, which is darker colored, red or purple on the back and orange below. It develops in a different way. The eggs are laid early in the summer and carried under the arms of the mother until the young are able to crawl away. Some of them, however, escape and may be sometimes seen at the surface as bright red specks, moving slowly by cilia all over them and gradually changing into the shape of Fig. 82. The young have this deep orange red color until they leave the parent.

Fig. 74.— A branch of *Bugula turrita*, showing the birds' heads.

The long-armed starfishes, Fig. 83, often come up above low water mark, under stones and in clusters of mussels, and are oftener found in the roots of large seaweeds thrown up on the shore. Instead of creeping by suckers they walk with their arms or squirm into cracks like a bunch of worms. These have free swimming

transparent larvæ of complicated shape similar to those of
the common starfish.

The common sea-egg, *Echinus*, Fig. 84, is common along

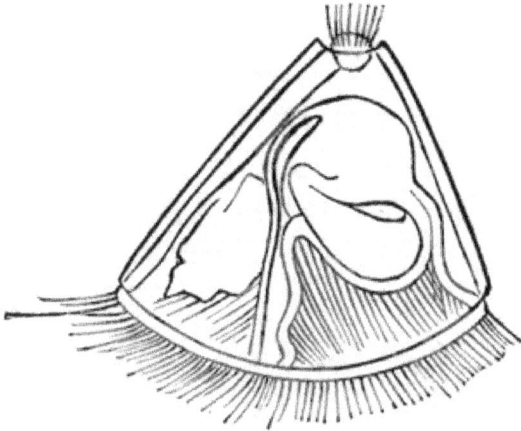

Fig. 75.—Young Polyzoa, *Menbranipora pilosa.* [From Barrois.]

rocky shores and small specimens may be found under stones
between tides, but the large full grown ones live where they

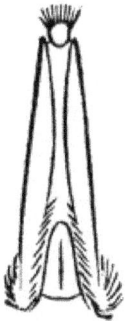

Fig. 76. — Swim-
ming young of
Polyzoa, *Mem-
branipora pi-
losa*, seen edge-
wise. [From
Barrois.]

are always covered with water and are sheltered in cracks or among small stones. They
may often be found by feeling under stones
just below low water. The resemblance of
one of these sea-eggs to a starfish is not
very evident at first sight of a specimen out
of water; but let one be put in a dish with
water enough to cover it, and it will soon
straighten up its spines and put out five double
rows of suckers, which extend around the
body from one pole to the other instead of
being on one side only as in the starfish. The
sea-egg may be compared to a thick starfish with the back very

small and the arms turned together upward. The suckers can
be run out a long distance and the sea-egg walks quite rapidly
with them. The greater part of the body is covered with
spines which are movable and appear to help the animal
along and to keep it clear of stones and other objects among
which it crawls. Each spine rests on a little knob and is
moved by muscles radiating around its base. Among the
spines are the "pedicellariæ," Fig. 85, of several kinds and
raised on long stems so as to reach beyond the ends of the
spines. They are three-jawed and, if any part of the animal

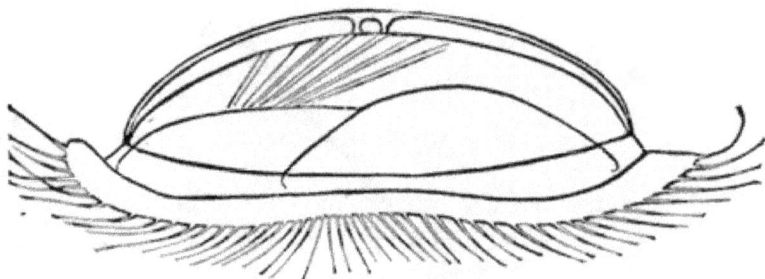

Fig. 77.—Young Polyzoa, *Flustrella hispida.* [From Barrois.]

is touched, turn toward it and open their jaws. Their office
seems to be to keep the surface of the shell clean by picking
up bits of dirt and passing them along where they will wash
off. The pedicellariæ can be seen in operation with a low
magnifying power, but to understand their shape and structure
it is necessary to pull some off and put them under a higher
magnifying power. The sea-egg lives on plants which it
gnaws with fine sharp teeth, with which its mouth is provided,
and which are supported by a complicated framework inside
the body. In confinement the sea-eggs need frequent changes

of water and are constantly discharging balls of excrement which help to make it dirty.

The porous plate which covers the entrance to the water system is found also in the sea-egg near the middle of the upper side, Fig. 86. The fine little holes in a circle around this end are the openings through which the eggs are dis-

Asterias.

Fig. 78.

charged. When the sea-eggs die they soon decay, and the spines drop off leaving a thin shell covered with little knobs, where the spines were, and rows of holes where the suckers came out. In this condition they are washed up on the beaches and are better known among collectors of curious things than in their natural condition.

Another animal allied to the sea-egg is the "sand dollar," Fig. 87, a flattened-out echinus, with a fur of little spines and very small suckers among them.   It lives on sandy bot-

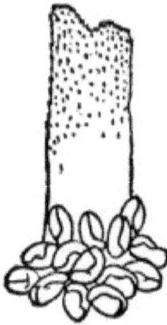

toms seldom above low water mark and partly buried in the sand. When living, its color is dark purple which soon fades to green and gray when it is taken out of the water. "The fishermen on the coast of Maine and New Brunswick sometimes prepare an indelible marking ink from these sand dollars by rubbing off the spines and skin, and after pulverizing making the mass into a thin paste with water."*

Fig. 79.—A spine of Starfish with a cluster of pedicellariæ around the base.

Related to these animals are the soft-bodied "Holothurians" one species of which, *Leptosynapta Girardii*, Fig. 88, lives buried in sand like a worm, for which it might easily be mistaken. It is white and translucent with fine opaque lines that show the division into five segments as in the sea-egg.   It is about six inches long and a quarter to half of an inch in diameter and keeps buried entirely in the sand in holes leading down from the surface. At the upper end is the mouth surrounded by ten soft tenta-cles. The intestine is usually full of sand from which the animal appears to extract its food. The skin is full of little hard plates with anchor-shaped hooks attached,

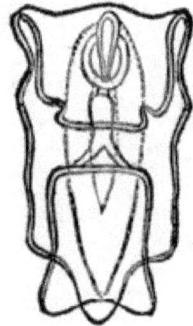

Fig. 80.—Young Star-fish. [From Agassiz.]

---

* Invertebrate Animals of Vineyard Sound.

Fig. 89, which make it stick to everything which touches it. Some other of these holothurians will be mentioned farther on among animals of deeper water, some of which have five rows of suckers and may be compared to an elongated sea-egg, while others creep on one side and look more like snails than they do like radiate animals.

The sea-anemones have long been favorite seaside pets both on account of the ease with which they can be kept in confinement and their supposed plant-like form and habits so different from all the more familiar animals. The common fringed sea-anemone, Fig. 90, lives among rocks near low water mark and still more abundantly on the piles of wharves and

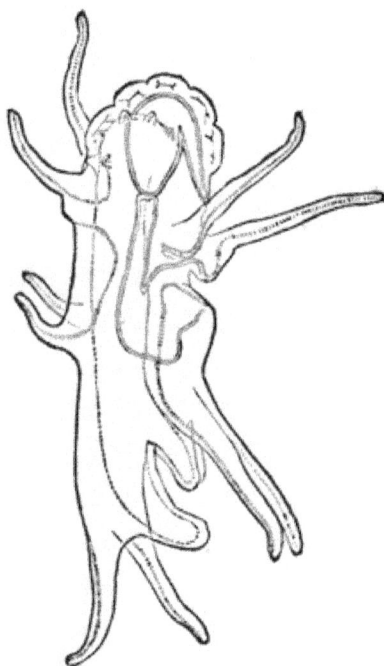

Fig. 81.—Young Starfish, showing at the upper end the fine lobes, which become the arms of the adult.

bridges, even where these are washed by a considerable quantity of brackish and dirty water at every tide. They vary much in size and color, but when contracted the greater part of them resemble in both these respects a baked apple. Other individuals are white, yellow, or spotted. The tentacles and the parts around the mouth which show when the animal expands are lighter, being flesh colored or various

shades of gray.   The outer tentacles are small and very
numerous, scattered over a disk which is scalloped and

Fig. 82.—Young
Cribrella.

folded at the edge.   Inside and around the
mouth are a few larger and longer tentacles
which, when expanded, stand up straight from
the disk. The edge of the mouth is wrinkled
and has two smooth grooves leading into it at
opposite sides.   The sea-anemone lives on
small animals of various kinds that come
against its tentacles as it stands spread out
in the water.   They adhere to the tentacles
and are passed along into the mouth.   The sea-anemone
sometimes runs out white
threads from its mouth or
from holes around its sides.
These threads are filled
with "nettle cells" which
can be conveniently seen
by pressing part of one of
these threads between two
pieces of glass under the
microscope.   The nettle
cells are large and long and
each contains a thread

Fig. 83.—Starfish, *Ophiopholis aculeata.*

coiled up within it which is thrown out when it is irritated ;
such cells are found on other parts of the body and are
common in all the animals of this class.   Fig. 91 is a section
across a sea-anemone to show how its body is divided by
partial partitions.

The sea-anemones though usually stationary are not permanently attached, but can crawl about slowly by the muscular base on which they stand. If kept in an aquarium they soon attach themselves and will creep up the glass where both top and bottom can be conveniently seen, and in small transparent individuals a considerable part of the internal organs. This sea-anemone is very easily kept in confinement especially in winter when the

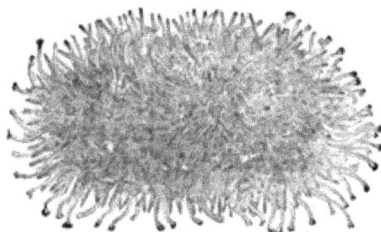

Fig. 84.—Sea-egg, *Echinus*, side view. [From Tenney's Zoölogy.]

aquarium can be easily kept cool. In summer the water must be changed or strained oftener to keep it cool and

Fig. 85.—Pedicellaria of Sea-egg.

clean, and the larger the amount of water the better. In keeping these and all other water animals, it is better to use small ones as these require less water and show almost everything that the larger individuals do.

On the southern coast of New England and farther south the "white armed sea-anemone, *Sagartia leucolena*, Fig. 92, is common near low water especially on the under side of large stones sometimes nearly buried in gravel. This is more elongated and more slender than the last and has a smaller, simple and plain disk,

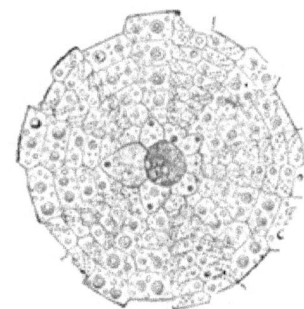

Fig. 86.—Centre of upper side of Sea-egg with spines rubbed off. [From Tenney's Zoölogy.]

with the tentacles much longer and more slender and

crowded together near the margin. The surface of the body
is usually pale salmon or flesh color and the skin is translucent so as to show the internal lamellæ; the tentacles are
paler and more translucent and usually whitish, but sometimes pale salmon. The tentacles in full expansion are over
an inch long."

The sea-anemones, although very much like the coral animals, make no coral themselves; but there is one species

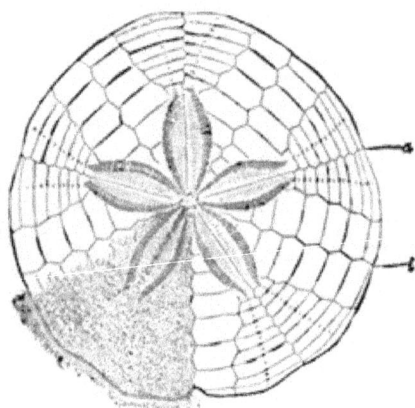

Echinarachnius parma.

Fig. 87.

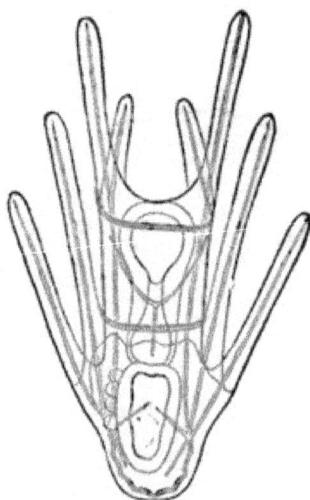

Fig. 87 a. Young of Echinus.

found on the southern coast of New England which makes
coral lumps of considerable size. "The *Astrangia Danæ,*
which is the only true coral yet discovered on the coast of
New England, is occasionally found on the under side of
overhanging rocks, or in pools where it is seldom or never
left dry. The coral forms incrusting patches, usually two or
three inches across and less than half an inch thick, composed of numerous crowded corallets having stellate cells
about an eighth of an inch in diameter. The living animals,

Fig. 93, are white, and in expansion rise high above the cells and expand a circle of long, slender, minutely-warted tentacles which have enlarged tips. These coral polyps, when expanded, resemble clusters of small, white sea-anemones, and like them they will seize their prey with their tentacles and transfer it to their mouths. They feed readily in confinement, upon fragments of mollusca and crustacea."* The coral formed by this polyp resembles, on a small scale, that formed in such large masses in more southern waters. After the animals are dead and their softer parts decayed, it is left as a stony lump covered with little rounded pits with

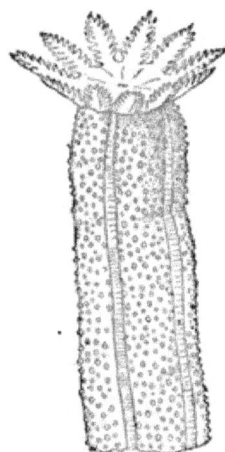

Fig. 88.—Upper end of a worm like Holothurian, *Leptosynapta Girardii*, with the tentacles partly extended.

radiating partitions extending from the circumference nearly to the centre. Each of these pits is the place where one of the polyps stood and has been secreted by it, as the shell

Fig. 89.—Plates and hooks from the skin of *Leptosynapta Girardii*.

of a snail or the bones of a quadruped are secreted, the partitions being formed in radiating folds such as may be seen in the base of a sea-anemone attached to a glass. All corals are formed in some such way and are not to be considered

as nests built by "insects" or other animals in which to hide themselves.

---

* Invertebrate Animals of Vineyard Sound.

Along with the fringed anemone there grow great bunches of a large Tubularia, *Thammocnidia spectabilis*, Fig. 94. It has a long transparent stem at the end of which is a circle of tentacles that spread sometimes over an inch in width. In the middle, above these, rise the stomach and proboscis with a circle of smaller tentacles around the mouth; at the end

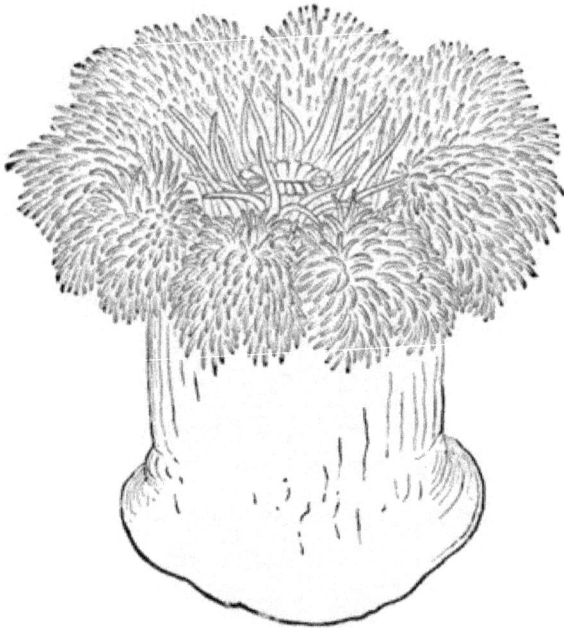

Fig. 90.—Common Sea-anemone, *Metridium marginatum*.

and between the proboscis and the outer tentacles are bunches of round objects like berries which are the ovaries, in which may be seen young polyps of various stages with their arms folded together as they become ready to hatch. These after a while drop off as little eight-armed polyps, which float about for a short time and then become attached by the under side and grow up like their parents.

Besides this method of increase, these tubularias also branch, and a new head and tentacles develop on the end of each of the branches. When they are kept in confinement the heads often drop off and after a time new ones grow on the same stems. In early summer, before the hottest weather, it is easy to keep this species and see the young fall off and attach themselves to the dish. Among the heads

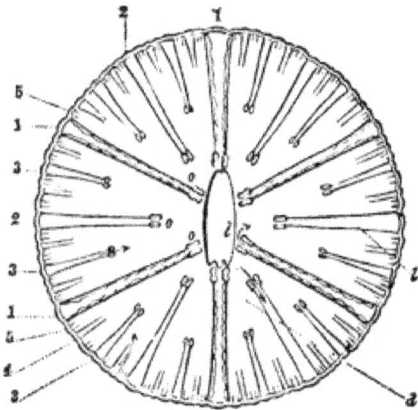

Fig. 91.—Section of Sea-anemone to show the radial partitions of the body.

that become detached from these tubularias are often great numbers of young *pycnogonidæ*, as will be described hereafter, which live there as parasites.

Clava leptostyla, Fig. 95, is another hardy polyp that grows near or even above low water. It is bright red and often lives on dark seaweed or mussel shells where

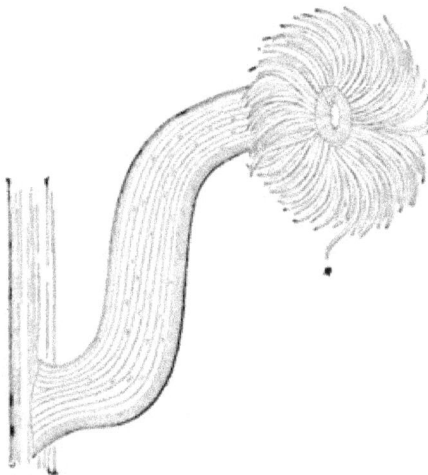

Fig. 92.—*Sagartia leucolena*. [Verrill,]

it shows itself very distinctly, lying down and partly drying at low tide but reviving when the water covers it. The tentacles

are arranged irregularly around the outer end, and in the
breeding season the ovaries are clustered below them as in
the figure. Most of the year, however, these are absent and
they have no branches except the tentacles.

Coryne mirabilis, Fig. 96, is another species something
like this. The buds below the tentacles, however, instead of
producing young directly grow into jelly-fishes like *d*, Fig. 97,
which become half an inch across and are among the most

common jelly-fishes.
They are very hardy
and can be kept for
some time in confine-
ment if the water is
clean.

Most of the plant-
like objects growing
around the shore are
not plants but Hy-
droids, compound
animals allied to

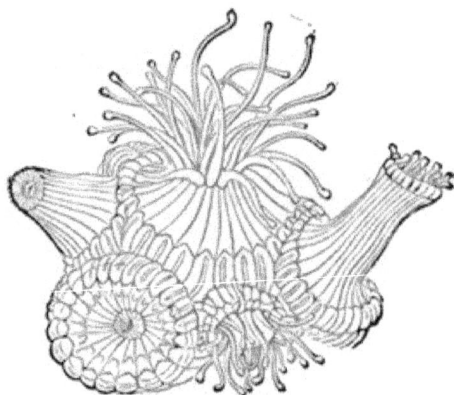

Fig. 93. — Five polyps of *Astrangia Danæ*. Upper
central one with tentacles fully expanded.

those just described, Fig. 98. They branch off originally
from one individual and grow into clusters of different shapes
according to the species. They look so much like plants
that almost every collection of seaweeds includes some of
them, and they keep very well pressed on paper like plants ;
but the whole of them, stems and all, belong to the animals
themselves whose mouths and tentacles are on the ends of
the branches, while the lower parts are more or less connected together through the stem. All of the branches,
however, do not have mouths and tentacles on the ends,

but instead have capsules which contain little buds which develop into ciliated young that are discharged and swim for a short time, and then grow up into new branching colonies like their parents ; or the buds develop into jelly-fishes that grow up and produce eggs, which in turn grow into stationary branching hydroids, Fig. 96. More examples of this "alternate genera-tion" of hydroids and jelly-fishes will be given in the chapter on swimming animals. The general shape of the clusters of hydroids is preserved well enough by pressing them on paper after the manner of plants ; but the soft parts are lost altogether in this way, and to keep them it

Fig. 94.—*Tham-mocnidia specta-bilis.*

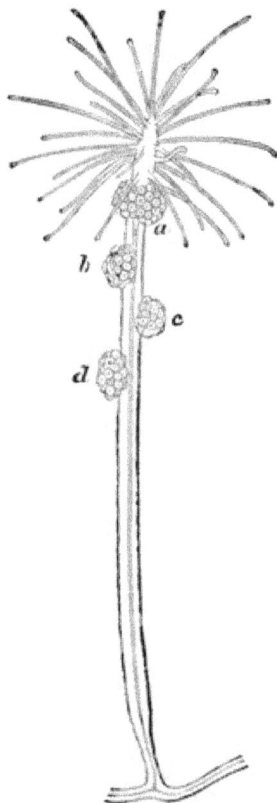

Fig. 95.—*Clava leptostyla,* enlarged. [After Agassiz.]

is best to use alcohol which preserves the polyps so that they can be examined very easily though con-tracted somewhat.

There are several sponges which grow around the shore, some even above low water. One of the commonest of these is *Ascortis fragilis*, a little white species that grows under stones among shells, often nearly covered up with mud. It consists of a number of little tubes about an inch high, connected together by other branching tubes at the base and when clean, which is seldom, nearly white and looking rough under the

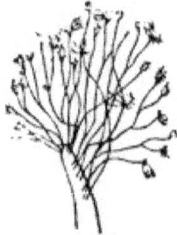

Fig. 96.—Cluster of *Coryne mirabilis*, natural size. [After Agassiz.]

microscope from the sharp spicules which fill them. To see these spicules, a small piece should be torn off and placed under the microscope and a little potash used to dis-

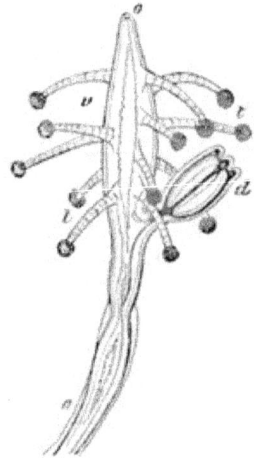

Fig. 97.—*Coryne mirabilis*, enlarged. [After Agassiz.]

solve the rest and leave the spicules clean. In sponges, there are little cavities lined with ciliated cells, that by their motions keep water passing over them and take in from it the

Fig. 98.—Cluster of *Sertularia pumila*. [After Agassiz.]

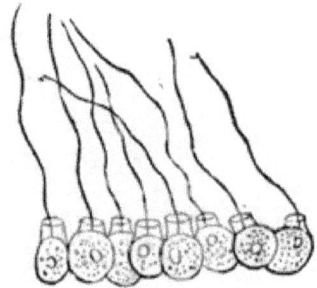

Fig. 99.—Ciliated cells of *Ascortis fragilis*.

sponge's food. These cells are very small, but loose ones may be seen when a piece of sponge is examined fresh. The ciliated cells (Fig. 99) of this species have been described and figured by Clark.

A larger sponge, *Chalina oculata*, Fig. 100, with a fibrous framework, is often found just below low water. It is generally attached by a small base from which finger-shaped branches spread into a half round cluster. This is more

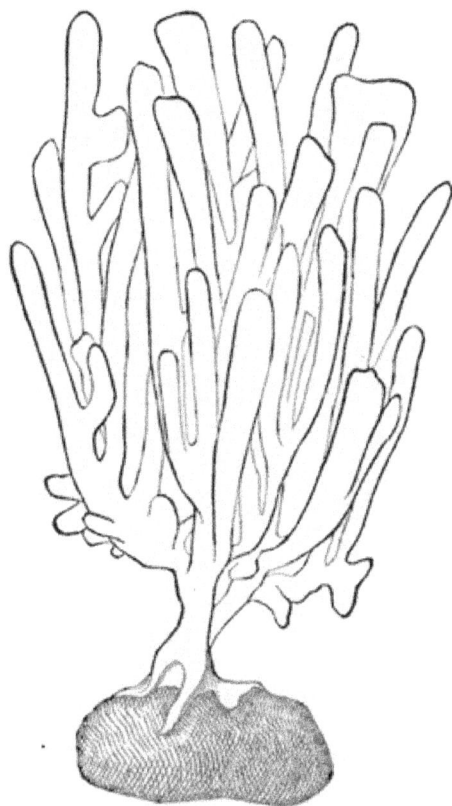

Fig. 100. — *Chalina oculata.*

like the washing sponges, but its fibres are too brittle for it to be of any use.

While examining animals with a microscope, infusoria of various kinds will often be met with, some swimming like the young of higher animals and others attached to shells and plants. The stems of hydroids are often covered with com-

Fig. 101. — *Vorticella* from hydroid stems

pound infusoria like Fig. 101, mixed in with the heads of the
hydroids themselves, so that to the naked
eye they may not be distinguished apart.

The stems of these animals are contractile and if one is touched the whole
cluster suddenly draws together into a
ball. When they are expanded a ring of
cilia is seen around the outer end that
keeps up a whirl of water around the
point where food enters. These currents
of water can be seen better if a little
indigo or carmine is put in the water,
and particles of the colored powder will
soon be swallowed and can be seen inside the infusoria. The cilia do not form
a circle, but a spiral with the centre near
the opening where the food is taken in.

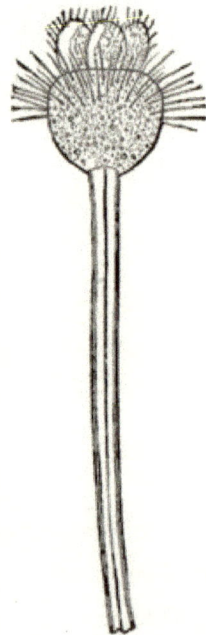

Fig. 102. — Acineta.

There are other similar infusoria that live singly each with a separate stem. Another kind of Protozoa, one of the Acinetæ, often found on hydroid stems, is shown in Fig. 102. Its stem is not contractile and the cilia on the outer end are straight and stiff. The lobe-like prominences on the top are young, which bud off in this way and finally drop and become attached by stems of their own. Other kinds have the cilia in two bunches and can contract and extend them.

SURFACE ANIMALS.

SURFACE ANIMALS.

WHILE the most interesting of the beach animals must be searched for among mud and stones, there lives another group that are always afloat in clean water and may be picked up with a dip-net or skimmed from the surface as the boat sails along. A few of these, such as the large jelly-fishes, can not fail to be seen by everybody who fishes or uses a boat. but most of them are very small and a large part of them nearly transparent, so that to the naked eye, looking from above, they are invisible. The number of animals at the surface is usually greatest at night and in suitable weather, especially on calm and warm nights, when they show their presence in a peculiar way, by the phosphorescence of the water as it is called. Sometimes this is so great that the ripples along the shore are enough to excite it and appear as white lines on the water, and at such times the boat and oars are surrounded by white light and drops from the latter look like sparks of fire. This phosphorescence often occurs on decaying fish; even on fish that have not long been dead the slime on the surface may shine in the night. Nearly all the small surface animals shine in the same way, though it is difficult to trace the light to individuals.

The best method of obtaining these small animals, where they are numerous, is to skim the surface with a muslin dip-net; but where they are scattered, as is usual in the daytime, a net may be arranged to tow after a boat, as in the picture, or to fasten where the current is strong and let the water run through it. The net should be taken up often so that the animals shall not be crushed against the cloth, and the inside washed off in a small quantity of water. To find the animals in the water, it should be placed in a glass vessel and a strong light thrown into it from the sun or a lamp, so that the transparent objects may be seen. A vessel with a black bottom, or a glass dish standing on a black surface, is the best to show such objects while they are lighted from above or from one side. Most swimming animals have the habit of collecting together toward the light, and a large proportion of those in the vessel will be thus gathered in a small space in a short time. Other less active species collect around the edge of the water and may be found by looking over this line with a magnifying glass. Some animals settle to the bottom when caught, but after a time swim up again; others are carried down by dirt adhering to them and can not rise, so that it is well to pour off the clear water and dilute the sediment with a fresh supply in order that the small animals mixed with it can be more easily picked out. For this work a pipette or medicine dropper with a rubber top is very useful, and a good supply of watch crystals, butter plates and other little dishes, should be kept on hand in which to separate them. After picking out everything visible, the water should be left standing quietly over night when new things will often come to the surface.

The first thing that will attract attention in a dish of surface water is the great number of minute crustacea which gather

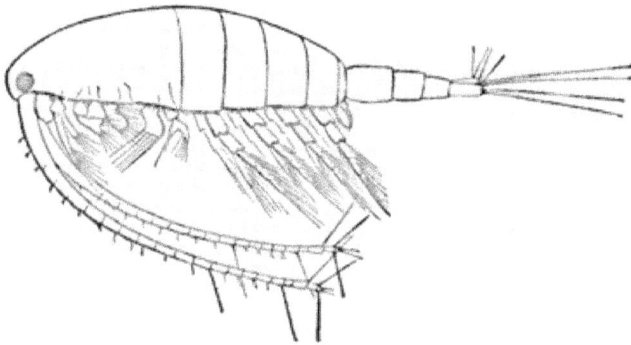

Fig. 103.—A Copepod.

toward the light. Some of these are to be found at all times of the year, several species occurring together on some days, while on others all will be of one species. It is useless to attempt to describe many of these little crustacea, for even among naturalists few of them are known; but one or two will be enough to illustrate the group. Fig. 103 shows one of these Copepods. One pair of antennæ are very long and sometimes modified into grasping organs in the male. The principal swimming organs are the short appendages under the tail. They appear to float at rest without effort and can then only be distinguished by the colored spots which most of them have in the head and antennæ.

When they do move, however, their motions are exceedingly quick and they dart here and there like a flash for a few moments and then settle quietly again. Some of them are almost always carrying their eggs in a bunch or two bunches behind, and these are usually more brightly colored than the animal itself. One of our largest species, half an inch long, carries its eggs in the spring. Its body is brown with a transparent spot behind the head, as though a piece had been bitten out, and the eggs are very bright green and carried behind the body attached to the hairs of the tail.

One of these copepods, "a species of *Sapphirina*, is one of the most brilliant creatures inhabiting the sea. It reflects the most gorgeous colors, blue, red, purple and green, like fire opal, although when seen in some positions, by transmitted light, it is colorless and almost transparent. When seen beneath the surface of the sea in large numbers, the appearance is very singular; for each one as it turns in the right position reflects a bright gleam of light of some brilliant color, and then immediately becomes invisible, and scintillations come from different directions and various these depths, many of them being much farther beneath the surface than any less brilliant object could be seen."*

The young of the common crab, Fig. 41, are found all summer long among these copepods, and are easily recognized by their comparatively slow motions and short jerks up and down along the edges of the water. With

---

* Invertebrate Animals of Vineyard Sound.

them occur also the young of hermit crabs and shrimps and occasionally of the lobster, Fig. 38.

The young of *Mysis*, Fig. 104, are also very common, distinguished by their large eyes extending each side of the head to twice the width of the body.

Some of the adult crabs swim at the surface as the common "blue crab," the "lady crab," *Platyonichus*, and the males of the oyster crab.

A curious little flat crustacean, *Caligus rapax*, is parasitic

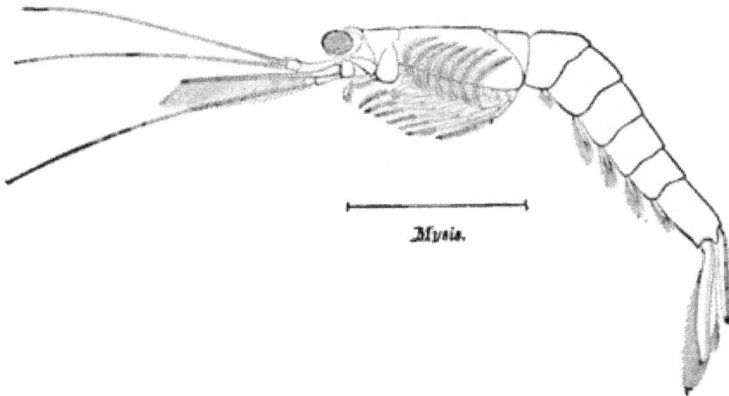

Mysis.

Fig 104. — *Mysis stenolepis.*

on fishes but swims much at the surface and is often caught in the net. It is from an eighth to a quarter of an inch long, dark brown in color, and the female often has two long egg sacs extending behind each side of the tail. They swim about rapidly in the dish when chased, but will sometimes hold so tightly to the sides that they can only be rubbed off.

Several species of Isopods and Amphipods are common at the surface, including some already mentioned as occurring under stones on the beach, as *Idotæa irrorata*, page 29, which is often taken swimming or holding to floating bits of

weed. *Idotæa robusta*, a slate colored species, with the tail
truncated, is usually found swimming farther out from the
shore and probably is afloat much of the time. A curious
amphipod, *Hyperia*, is also often taken. It has a short body
and a head flattened in front with short antennæ and immense eyes. It is parasitic on jelly-fishes. Among these
crustacea may be mentioned the cast-off skins, especially of
barnacles, which are always floating and easily mistaken for
living animals.

On the bottom of the dish are usually some small roundish
crustacea with slender forked tails with which they push them-

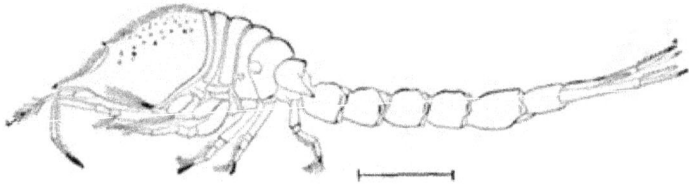

Fig. 105.—*Diastylis quadrispinosus*, straightened out.

selves awkwardly about among the dirt, but now and then one
will turn the tail close over the back so as to look like a seed
and swim up to the surface. These belong to a group called
the Cumacea and there are several common species, Fig.
105.

The large sea worms, Nereis, swim at the surface especially
at night and in the breeding season. The males are, however,
oftener found at the surface than the other sex. The adults
of the smaller species of Autolytus swim all the year round.
They are small worms half an inch to an inch long with long
bristles at the sides and long curled appendages on the head.

The males, Fig. 106, may be distinguished by six or more narrow segments just behind the head, and the greater length and size of the appendages on the other segments and on the head. The females usually carry a bag of brightly colored eggs under the middle of the body, of different colors in different worms. The young may be seen through the sac as minute triangular worms each with two red eyes. The young do not swim like the adults, but live in tubes among plants and hydroids and as they grow up divide into several worms fastened head to tail. The individuals that drop off behind are those that develop into free swimming adult males and females.

A great many worms, which burrow in the mud when adult, swim at the surface when young, and great numbers of these little larvæ, Fig. 107,

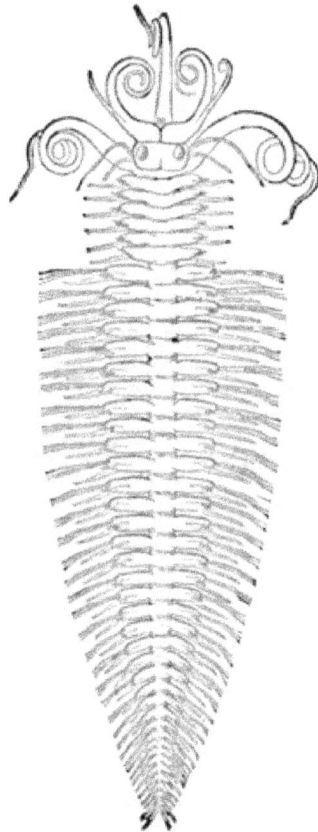

Fig. 106.—Male *Autolytus.*

may be found in the net some swimming by a ring of cilia around the head, and others a little farther advanced with a few bristles and paddles often longer than those of the adults.

Some worms with smooth bodies or with very short appendages manage to swim at times by wriggling the body spirally, and others with flatter bodies by a waving up and down motion like a leech.

In the winter and spring the water is often full of straight transparent objects that usually lie still and stiff, but now and then dart across the dish and come to rest again. This is *Sagitta elegans*, a transparent worm. The middle of the body is smooth with two pairs of stiff fins at the sides and a fin at the tail like that of a young fish, but crossing the body horizontally instead of up and down. The head has a pair of eyes and three pairs of bunches of bristles closed together near the mouth.

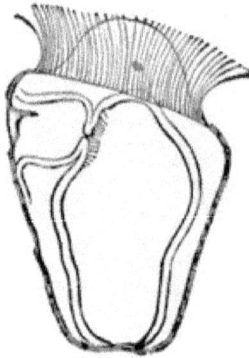

Fig. 107.—Young worm.

Although in this volume the fishes and other vertebrates have received but little attention, we can not omit to mention the young fishes and fish eggs that form so large a part of the surface fauna, especially in the spring and summer. The young of many fishes, which when full grown feed usually at the bottom, swim at the surface until an inch or two long, when they disappear and are hard to find until large enough to bite the hook. The young of the grubby, *Cottus*, of the lump-fish, *Cyclopterus*, and of the cod and hake swim near the surface in this way. In like manner the eggs of many fishes float at the surface, as for instance those of the cod. They are very small and transparent so that they are invisible except in the most favorable light, but a line of them will often collect around the edge of the water and they can then be seen with the naked eye, or better with a magnifying glass of low power. The development in these eggs goes on rapidly and those which appear perfectly clear when taken will sometimes show the next day a very fish-like embryo. The young fishes hatch in a very immature state with a large ball of yolk still attached under them, and may often be taken at the surface in this condition, so transparent that they could hardly be found except by their color marks.

Among the largest swimming animals are the squids or cuttle-fishes, Fig. 108. The common species are well known to fishermen, by whom they are caught for bait, and are about a foot in length with a distinct head with large and bright eyes, in

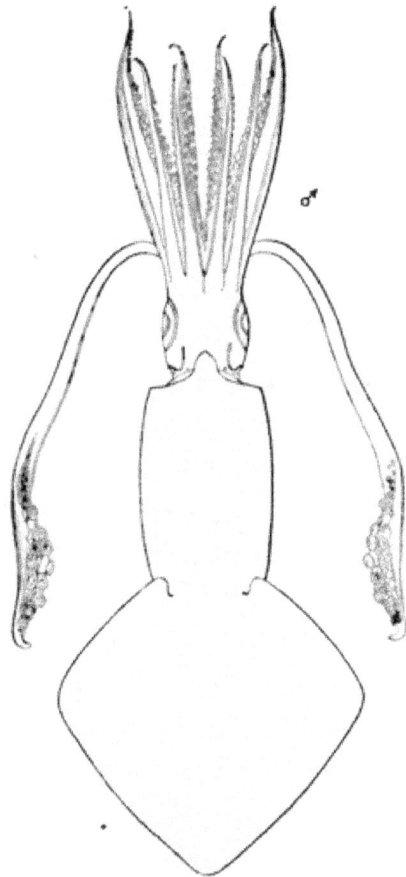

Fig. 108.—Squid, *Loligo pallida*. One-third natural size.

front of which are the ten strong arms covered on the inner side with suckers. The mouth is provided with a strong beak, like that of a bird, and has within a tongue covered with rasping teeth like the tongues of snails. The food is held by the arms and bitten off and scraped to pieces before being swallowed. The method of swimming practised by the squid is a very curious one. On the under side is a large chamber, nearly as large as the rest of the body, which contains the gills and the water which surrounds them. The opening to this chamber is by a slit just back of the head which can be closed when the animal wishes. In the middle of this slit is a flexible tube which can also be closed. When the animal wishes to swim it closes the sides of this slit and forces out the water through the tube, driving itself backward like a rocket. The direction can be changed by turning the tube and it can even shoot itself forward by pointing the tube backward. In the "Invertebrate Animals of Vineyard Sound," Smith and Harger give the following account of the habits of the squid, *Ommastrephes illecebrosa*, at Provincetown, Mass., which they watched along the wharves, July 28, engaged in capturing and devouring the young mackerel which were swimming about in schools, and at that time were four or five inches long. "In attacking the mackerel they would suddenly dart backward among the fish with the velocity of an arrow, and as suddenly turn obliquely to the right or left and seize a fish, which was almost instantly killed by a bite in the back of the neck with the sharp beaks. The bite was always made in the same place, cutting out a triangular piece of flesh, and was deep enough to penetrate to the spinal cord. The attacks were not always successful, and were sometimes repeated a dozen times before one of these active and wary fishes could be caught. Sometimes after making several unsuccessful attempts, one of the squids would suddenly drop to the bottom, and, resting on the sand, would change its color to that of the sand so as to be almost invisible. In this way it would wait until the fishes came back, and when they were swimming close to or over the ambuscade, the squid, by a sudden dart, would be pretty sure to secure a fish. Ordinarily, when swimming they were thickly spotted with red and brown, but when darting among the mackerel they appear translucent and pale. The mackerel, however, seemed to have learned that the shallow water is safest for them and would hug the shore as closely as possible, so that in pursuing them many of the squids became stranded and perished by hundreds, for when they once touch

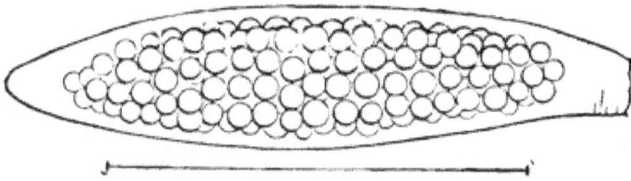

Fig. 109.—Cluster of eggs of *Loligo Pealii*.

the shore they begin to pump water from their siphons with great energy, and this usually forces them farther and farther up the beach."

The changes of color which they undergo to make themselves more like their surroundings are caused by changes in the size of their color spots, which are constantly taking place and can be best observed in young specimens.

The squids have one very peculiar habit. Inside the tube, through which the water is discharged from the gill chamber, is a bag in which ink is secreted and when the squid is anxious to escape, it discharges some of it into the water, which is instantly blackened so that the squid can not be seen.

The eyes of squids are large and they are to some extent nocturnal in their habits and will collect around a light so that they can be driven ashore in large numbers.

The eggs of one of the squids, *Loligo Pealii*, are laid in clusters enclosed in jelly, and stuck together in large masses which are sometimes washed up on the shore, Fig. 109.

The body of the squid is stiffened by an elastic piece called the pen, Fig. 110, which is in the skin of the back and is of various shapes in different kinds of squids. The cuttle-fish bone used for canary birds is the pen of a short-bodied cuttle-fish.

Farther out to sea there are larger squids which are occasionally found in the stomachs of fishes or whales, or thrown ashore dead or disabled. Within the last few years several specimens of gigantic squids have been cast ashore at Newfoundland from which measurements have been taken and parts of them preserved.

One of the most complete of these was taken whole to New York and is preserved in a somewhat damaged condition at the New York Aquarium. When fresh this specimen measured nine and one-half feet from the tip of the tail to the base of the arms and was seven feet in circumference. The length of each of the long tentacular arms was thirty feet and of the longest of the other arms eleven feet. The largest suckers were an inch in diameter and had a row of sharp teeth around the edge.

Fig. 110.—Pen of a Squid, *Loligo pallida.*

Fig. 111 is copied from a restored drawing of this specimen

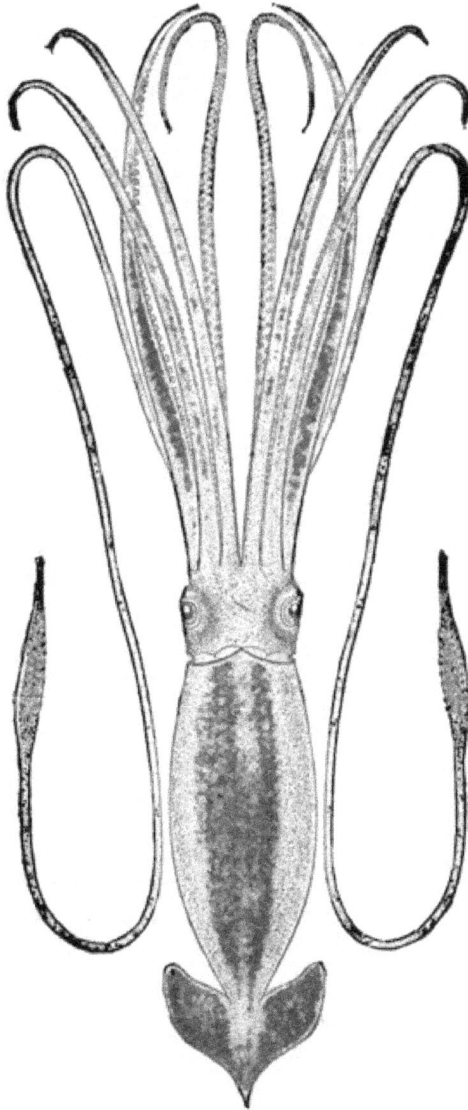

Fig. 111.—A large squid, *Architeuthis princeps*, one-fiftieth natural size. [Restored from the specimen in New York by Prof. Verrill.]

by Prof. Verrill, lately published in the Proceedings of the Connecticut Academy of Sciences.

It can easily be imagined that such creatures as these could drown a man, or upset a boat, and they have furnished material for many fabulous accounts of sea-serpents and devilfishes. These large squids are probably more numerous than was formerly supposed, as since the recent finding of several thrown up on the shore, numerous fishermen have reported seeing them near the fishing banks and have, in some cases, caught and used them for bait. Pieces of them sound enough to be recognized have often been seen by whalemen, thrown up from the stomach of whales when dying, and some of the largest jaws of squids known have been taken in this way.

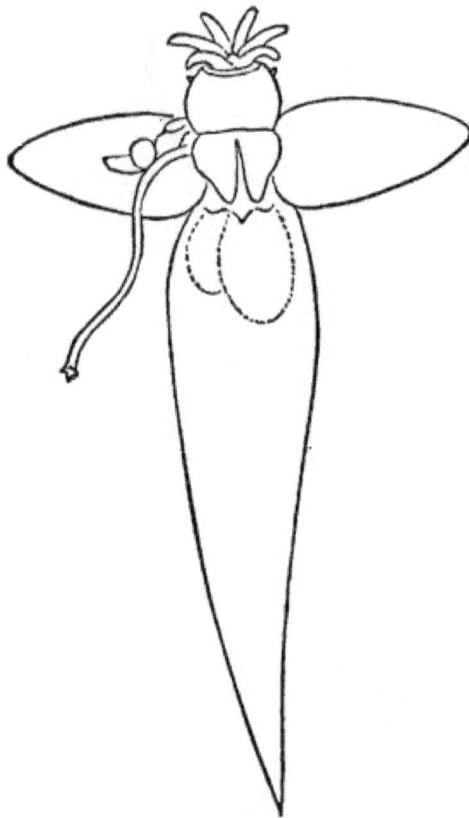

Fig. 112. — *Clione papillonacea.*

The Pteropods, Figs. 112 and 113, are another group of surface animals which swim by two wing-like appendages just behind the head. Fig. 112 is one of these, *Clione papillonacea*, from a sketch by Prof. Verrill. The mouth is at the upper end surrounded by six tentacles. The short appendages between the wings correspond to the "foot" of snails, and the apparatus at the side belongs to the reproductive organs. This species is not common, living usually at some distance from land. Other Pteropods have shells, that of *Styliola vitrea* being a long cone, of *Spirialis* a spiral, and of *Cavolina tridentata*, Fig. 113, a complicated shape. They all, however, swim in the same way by a pair of wings which are extended out of the shell.

Many snails swim at the surface when young by means of cilia on two wing-like appendages similar to those of Ptero-pods, but which disappear as the snail grows up. These temporary appendages are nowhere more distinct than in the young of the naked mollusks which lose not only these but the shell, when they become adult. This is best observed in the young of *Eolis diversa*. Over the lower part of the mouth of the shell extends the foot, covered

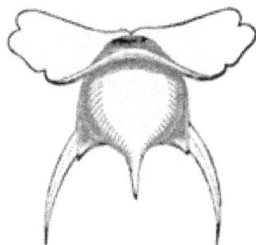

Fig. 113.—Pteropod, *Cavolina tridentata*.

with cilia and carrying the flat "operculum" which, when the animal is entirely contracted, closes the shell. Above are the lobes which have around their edges long cilia by which the young snail swims through the water. The large round objects seen inside are the ears. The eyes are on top of the head behind the swimming lobes. They swim much of the time when first hatched, but later rest oftener on the bottom. They can be easily raised from the eggs and kept a short time living in confinement. These Eolis young are almost transparent, but other species are brightly colored, usually around the edges of the swimming lobes just inside the large cilia, and can be easily seen in the water by these markings,

Fig. 114. They are apt to settle to the bottom when first caught and become mixed up with the sediment, but rise after the water becomes quiet. They are less likely to gather around the edges of the dish than most surface animals and swim up here and there all over it.

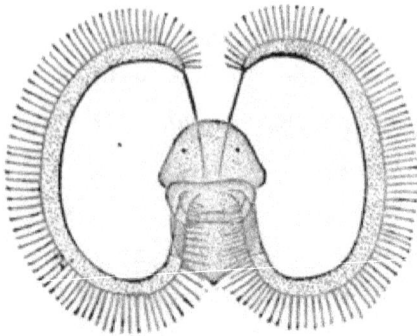

Fig. 114.—Veliger of unknown snail.

Most Ascidians are stationary when adult but swim free for a short time after hatching. Fig. 67 is the young of *Molgula*, a common ascidian under stones and among mussels. After swimming about for a time, the larvæ attach themselves to the bottom, secrete a thick covering over them, lose the tail and transform into stationary animals.

There are, however, other ascidians which always swim or float, the *Salpa*, Figs. 115 and 116. They are as transparent as jelly-fishes, sometimes slightly reddish with blue around the edges of the mantle and about an inch long. They swim by forcing out water from the posterior respiratory opening. The white lines

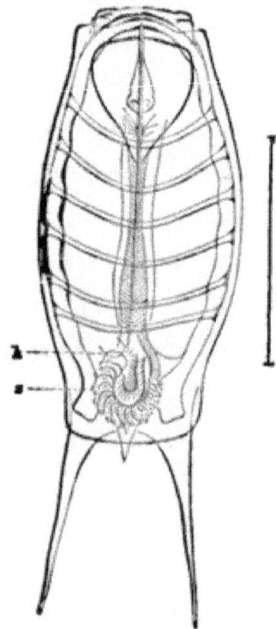

Salpa Cabotti Des.

Fig. 115.—*Salpa*, solitary individual with a chain of young, *s*, forming within it.

running around the body are muscles. When they are
mature there grows near the posterior end a chain of
little Salpæ united together in a double row of twenty
or thirty pairs. As the chain grows larger it is discharged into the water and the Salpæ grow up still connected together till the chain is a foot or more in length.
The whole moves by the discharge of water from the bran-
chial openings of the individuals
composing it. Each member of
these chains produces a single
egg which develops into a solitary individual that in its turn
produces another chain and so
on alternately. They grow rapidly and sometimes the water is
completely filled with them.

The young of starfishes and
sea-eggs have already been mentioned (pp. 61 and 63). They
swim deep during the day but
rise to the surface in the even-
ing and get into the net sometimes in considerable numbers.

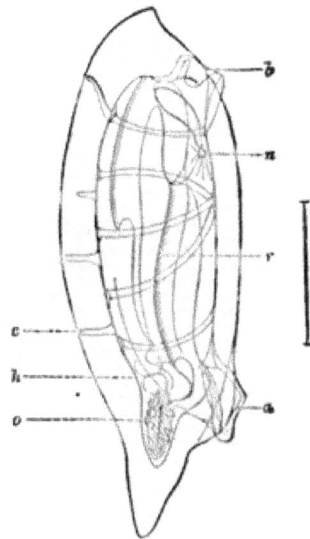

Salpa Cabotti Des.

Fig. 116.—*Salpa*, an individual from
a mature chain; *c*, the processes
by which they are held together.

They can, however, be raised artificially by opening a mature
female and discharging the eggs into water and then opening
a male into the same water and after stirring, changing the
water until no floating bits are left to decay. If they be
kept cool and the water changed daily the development
will go on regularly and can be watched from day to day.
As soon as the young begin to swim, the water can be

changed by drawing it off below them with a siphon and they can be raised in this way as long as they continue to float.

In the early part of the summer, little round spots of a very deep orange-red color are found at the surface. They move slowly by means of cilia all over them, but later have a circle of five tentacles at one end and two at the other, Fig. 117. These are the young of a large holothurian, *Lophothuria*, which will be described further on. Later in the season these red larvæ settle to the bottom and may be found among stones at low water with the tentacles branched and scales on one side like the adult. Along with these round red larvæ occur the similarly colored larvæ of *Cribrella*, Fig. 82.

Fig. 117. — Young Lophothurian.

Perhaps the most curious animals in the surface fauna are the jelly-fishes; some of them as transparent as water, and containing so large a proportion of it that, when dried, there is hardly substance enough left to show where they were. The larger species are familiar objects all along the coast in the early summer. Beginning in February and March as little disks not more than a quarter of an inch across, in the course of the season they reach a diameter of a foot or more and swim out several miles from the shore; and, before autumn, having dropped their eggs, they become opaque and dilapidated and are thrown up here and there on the beach. The most interesting part of their history, however, is that in their earliest stages most of them pass through complicated changes between the egg and adult.

To begin with one of the best known let us trace the growth of the large white jelly-fish, *Aurelia*, Fig. 118. In this jelly-fish are four colored masses half-way between the

Fig. 118.—Common white jelly-fish, *Aurelia flavidula.*

mouth and the rim which consist of eggs. Late in the summer, if the animal is put in a pan of water some of these become loose and slowly creep about on the bottom.

If examined closely, they are seen to be covered with cilia and it is by these that their motions are kept up. These eggs, or larvæ as they may now be called, are dropped loose in the water and those which happen to find suitable resting places become attached by one end and soon open a mouth at the other, surrounded by tentacles something like a minute sea-anemone,

Fig. 119.— Young of the white jelly-fish, *Aurelia.* First stage.

Fig. 119. They live in this way all winter, and some of them longer, for they may be found under stones in the

early summer. In the spring these polyps elongate and
divide transversely into a pile of jelly-fishes, Fig. 120,
which drop off one by one and swim away, Fig. 121. At
this stage they do not look much like the adult, but have
eight long arms forked at the ends and between these as
many rounded lobes. It is by no means rare, however, to
find one with a larger number of branches, twelve being a
common number. As they grow larger the lobes increase

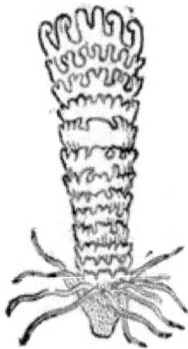

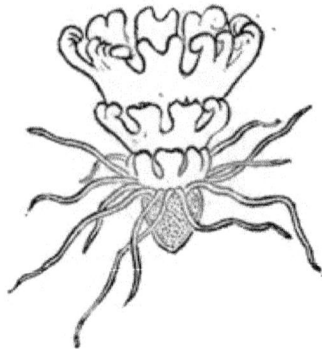

Fig. 120.   Fig. 121.— Young jelly-fishes ready to
                      separate and swim away.

more rapidly than the arms, so that the outline of the animal
soon becomes a circle slightly scolloped. The forked arms
carry between the prongs the eyes of the jelly-fish and these
remain in the adult. As the lobes grow, they develop along
the outer edge a fringe of five tentacles ; and at the same
time four large tentacles grow around the mouth, and the
ovaries begin as four clusters of tubes. The radiating
branched lines running from centre to circumference (see
Fig. 118) are tubes which branch from the central stomach
and run to a circular tube around the outer edge of the
animal and are the only circulating vessels that it has. It

swims from the first by contracting itself around the edge suddenly and then slowly spreading again for another stroke ; but it moves slowly and does not appear able to go far in a definite direction but to drift about with the currents, moving only enough to circulate the water around it and to change the depth at which it floats.

The red jelly-fish, *Cyanea arctica*, develops in much the same way, but it grows much larger and floats in deeper water. The tentacles which hang from the edge are longer and much more numerous, trailing out behind sometimes for ten feet from the disk. These tentacles have on them very strong nettle cells, and persons bathing are often stung by them so that their skin feels as if burnt. The white jelly-fish will sting the skins of some persons, even the hands, in the same way, but is harmless to most persons.

In spite of their stings, these jelly-fishes furnish shelter for

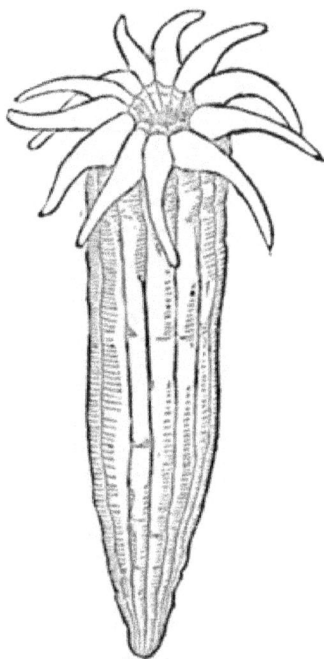

Fig. 122.—*Peachia parasitica.*

some other swimming animals. Certain little fishes hide under them and swim in and out at pleasure, though occasionally one gets caught and swallowed.

An amphipod crustacean, *Hyperia*, lives on them and there is a species of sea-anemone, *Peachia parasitica*, Fi

122, that lives in the pockets around the mouth and should be looked for whenever one of these jelly-fishes is captured. It is impossible to preserve these large jelly-fishes satisfactorily, but small ones under an inch in diameter can be kept very well in alcohol which should be used at first weak and gradually strengthened, so that the animal will contract without becoming distorted. As is usual, however, in such work,

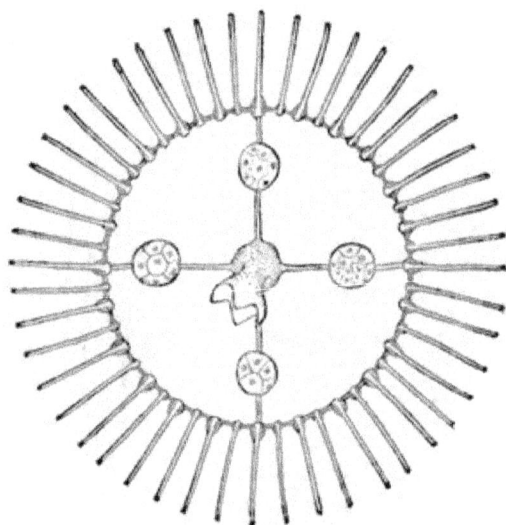

Fig. 123. — *Obelia.*

several specimens will be spoiled for every good one.

The development of other jelly-fishes goes on in a similar complicated way. In the water from the surface net, there are almost always some little flat, and very active jelly-fishes, *Obelia*, Fig. 123. They are usually about an eighth of an inch in diameter and are very transparent so that they are not easily seen until they move. Instead of the branching radiating tubes of the large jelly-fishes, they have four simple

ones connecting the stomach with a circular tube around
the edge. The stomach hangs down in the centre and has
a four-lobed mouth. The outer edge of the animal has a
fringe of tentacles spreading out nearly flat, and at the bases
of some of these are eyes. The four round bodies hanging
from the four radiating tubes are the ovaries usually filled with
eggs in various stages of development. These eggs grow up
into branching hydroids, Fig. 124, with a
mouth and tentacles on the end of each
branch, and form a part of the feathery
growths so common just below low water
mark. All the branches, however, do not
produce mouths and tentacles; but some of
them form long capsules filled with little
round bodies that gradually grow into jelly-
fishes, and are discharged into the water
where they grow up and lay the eggs for
another generation.

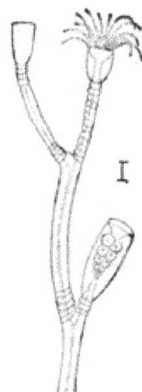

Fig. 124.—*Obelia
commissuralis.*
One of the upper
branches has the
tentacles fully
expanded. The
lower branch is
filled with buds
which grow into
jelly-fishes.

Another method of growth is that of *Coryne
mirabilis*, Fig. 125, which is one of the most
common jelly-fishes. It has the same four
radiating tubes as those just described, but
only four tentacles, one at the end of each tube with an eye
at the base of each tentacle. The stomach is very long, extending out of the bell, and small crustacea may often be
seen in it in process of digestion. This species swims more
strongly than those with flat disks. The eggs form around
the stomach and after they are laid grow up into stationary
animals (see Figs. 97 and 98), with a mouth at the tip and
tentacles along the sides for some distance below it. Among

these tentacles, buds grow out after a time which take the shape of little jelly-fishes fastened by the top (see Fig. 97), and these finally drop off and swim away.

One of the commonest jelly-fishes in early spring is the

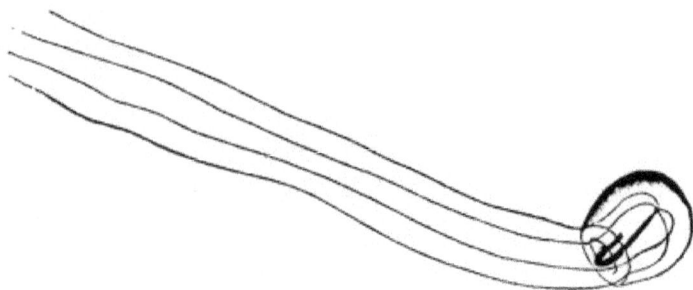

Fig. 125. — *Coryne mirabilis*, with the tentacles extended natural size.
[From Tenney's Zoology.]

Tiaropsis diademata, Fig. 126. It is very transparent and grows to be about the size of the figure. The proboscis is

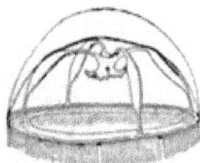

Fig. 126. — *Tiaropsis diademata.*

dark colored and scolloped and folded at the edges. The eggs are also dark and develop on the four radiating tubes. The young of this species are more deeply bell-shaped than the adult. They are very abundant in the winter and may be mistaken for another species.

In the late autumn a similar species, *Oceania languida*, becomes very abundant. It is even more transparent and

Fig 127. — *Tima formosa.*

has a flatter disk and fewer tentacles. It is often found with young fishes in the stomach.

But perhaps the most beautiful jelly-fish of this coast is
Tima formosa, Fig. 127. It grows to be two or three inches
across and is trans-
parent throughout,
except the four
folded ovaries which
run along inside the
bell from the edge

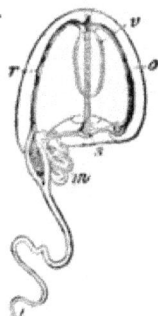

Fig. 128. — *Hybo-
codon prolifer*,
an unsymmetrical
jelly-fish with
buds at the base
of the large tenta-
cles. [From
Agassiz.]

to the mouth.

Besides the com-
plicated methods of
growth already de-
scribed, some of the

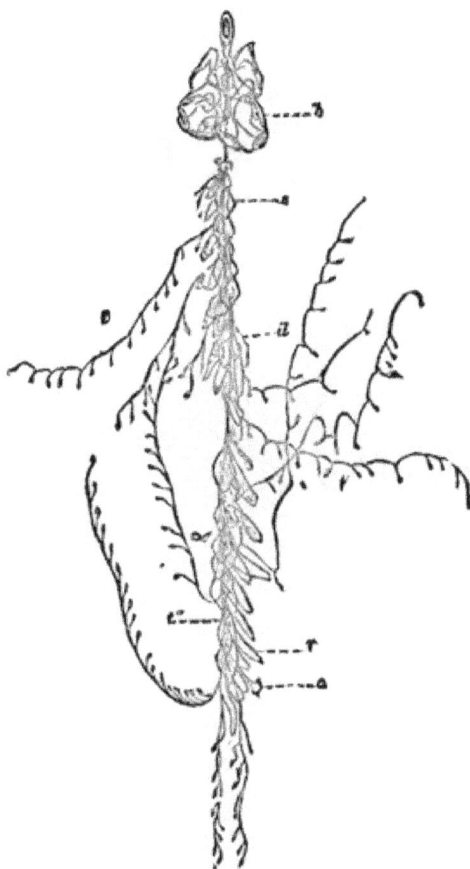

Fig. 129. — Compound jelly-fish, *Nanomia cara*.

jelly-fishes have the habit of budding, forming clusters of
individuals which, as they mature, break off and swim away,
Lizzia grata produces buds along the sides of the stomach
which grow into perfect jelly-fishes almost as large as the
parent before dropping off.

Hybocodon, Fig. 128, is one-sided with tentacles on only

one side.   The buds grow around the bases of these tentacles
and form bunches of young, some of which remain attached
until they are as large as the parent.

There are some compound jelly-fishes that form clusters of
many individuals of different
kinds, some catching the food
and eating it for the whole
colony, others moving the
whole through the water and
others still producing the eggs.
One of these, *Nanomia cara,*
is shown in Fig. 129. Another
famous compound jelly-fish is
the " Portuguese Man-of-war,"
Physalia, Fig. 130, which is
occasionally found along the
southern coast of New England,
but is more common farther
south.

The animal consists of a
large bladder-like float, under
which hang clusters of individuals of various kinds which·
carry the tentacles and diges-
tive and reproductive organs.
In life, the float is brightly
colored with red and blue.
The tentacles have the same
stinging property as those of the common jelly-fishes.

Fig. 130. — Portuguese Man-of-war,
Physalia arethusa, a compound jelly-
fish. [From Tenney's Zoology.]

There is another group of jelly-fishes, the *Ctenophoræ,* that develop in a more simple way than the Hydroids and swim during their whole lives. They may be distinguished by the eight rows of paddles running along their sides, which are

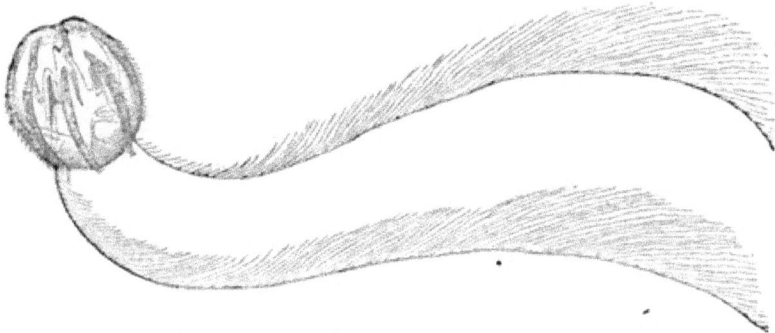

Fig. 131. — *Pleurobrachia rhododactyla.* [From Tenney's Zoology.]

in constant motion and reflect rainbow colors as the animal turns in the water. One of the commonest of these is *Pleurobrachia*, Fig. 131. It is nearly spherical and the rows of

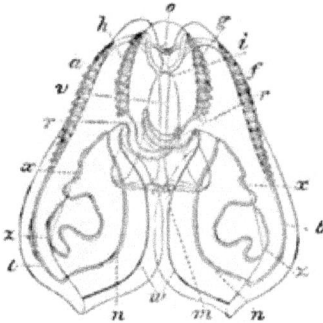

Fig. 132. — *Bolina alata.*
[From Agassiz.]

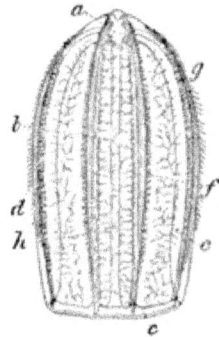

Fig. 133. — *Idyia roseola,*
reduced. [From Agassiz.]

paddles extend nearly around it. There are also two tentacles which can be contracted almost out of sight, but may be extended to several times the diameter of the body and spread out feather-like branches along their whole length.

Bolina alata, Fig. 132, is another species. It is differently shaped, having four large lobes around the mouth that make it appear wider at one end than the other. It often gets into the surface net but seldom comes out in good condition as it easily sticks to the cloth. The best way to procure these jelly-fishes is to watch for them and draw them near the surface with a net and then dip them up without taking them out of the water. They float at various distances below the surface, but can be seen against the dark water at a considerable depth. There is one species, *Idyia roseola*, Fig. 133, which is of a bright pink color. It is barrel-shaped and has no tentacles and the whole lower end of the body opens into the stomach.

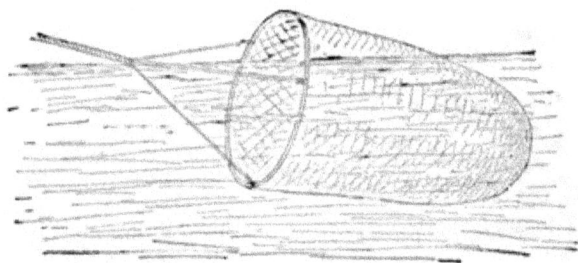

BOTTOM ANIMALS.

BOTTOM ANIMALS.

To the inland student nothing is so interesting as dredging for the animals that live beyond the lowest tides, out of reach of ordinary fishing and digging. Here he works upon a ground, only a small part of which has ever been explored and where he is sure to find something that he has never seen before. Aside from its scientific interest, dredging is a far more attractive employment than fishing for men who like active and out-door exercise with something to show for it afterward. The instrument most generally useful is the rectangular dredge, Fig. 134, with two scrapers, so that it will work whichever side falls on the bottom.

The frame of such a dredge can be made by any black-smith. The pieces at the sides are sometimes riveted into

the scrapers, but are better welded on or bent from the same
piece. The scrapers should not flare much. If they are an
inch farther apart at the edge than at the back, they will
scrape the bottom well enough and are not so likely to catch
in rocks or to fill the bag with mud as if they spread more.
For use by one person, or with a small row boat, a light dredge
made of hoop iron may be used; but for most purposes a
heavier one with scrapers half an inch thick and the sides
an inch or more in diameter, like that in the figure, is none
too stout for strength or weight. A row of holes should be

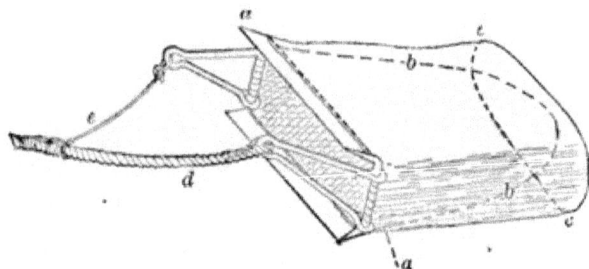

Fig. 134. — Dredge: *a*, iron frame; *b*, net; *c*, canvas cover; *d*, rope fastened
to one of the handles; *e*, small rope tied to the other handle.

made at the back of each scraper for attaching the net by
twine or, better, copper wire. It should also be tied to the
sides of the frame between the attachments of the handles.
Dredge nets suitable for frames, eighteen by six inches, or a
few inches larger or smaller, are kept on hand by the American Net and Twine Company, Boston, and will be tarred or
tanned if desired to protect them from decay or mildew.
To prevent the net from tearing or rubbing on the bottom
when full, it should have a cover of canvas put around it
open at the bottom and fastened to the dredge frame at the
forward edge. Leather is sometimes used for this purpose,

but it is more expensive and less easy to handle. It is a good plan when going on a dredging trip to carry an extra net and twine, needles and canvas for mending, so that if anything should become torn, the time need not be lost for want of a dredge. The way to attach the rope is shown in the figure. The end is tied to one of the handles, and to the other handle is tied a small line which is fastened to the dredge rope farther forward. In case the dredge is

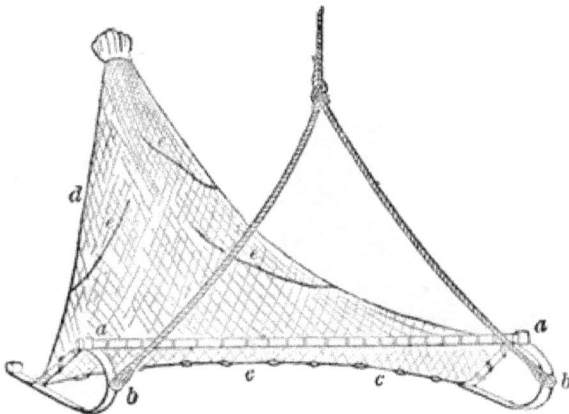

Fig. 135. — Trawl: *a*, beam which keeps the mouth of the net open: *b*, runners: *c*, lead line that drags along the bottom: *d*, net tied up at the bottom: *e*, pockets to prevent fishes from getting out.

caught among rocks, as often happens, the small line breaks and it comes up sidewise by the opposite handle. A weight equal to that of the dredge, or heavier, should be attached to the rope five or six feet in front of it, to keep the mouth turned down far enough to scrape the bottom. A dredge of this kind can be used anywhere. It is always safe before using other apparatus to take a haul with the dredge and see what there is on the bottom. Persons experienced in sounding can tell very nearly what the bottom is by the feeling of

the line as the lead strikes, and a small sample of the bottom
can sometimes be brought up by some tallow put in the hollow at the bottom
of the lead. With
a boat large enough
and a smooth bot-
tom, a "trawl" may
be used, Fig. 135.
This has nothing to
do with the trawls
of cod-fishermen,
but is a large net
with the mouth sup-
ported by a frame
resting on two iron
runners. The edge
of the mouth which
drags on the bottom
is weighted with lead
enough to keep it
down without dig-
ging into the mud.

The frame is some-
times so made that
the net is only at-
tached at the sides,
and both edges have
weighted ropes so
that it will work whichever side lies on the bottom. The
trawl may be made of any size that the boat and crew can

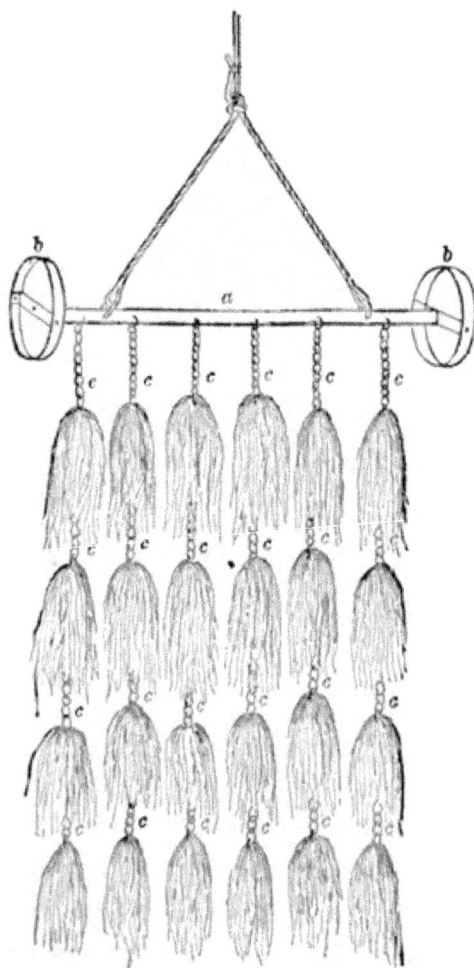

Fig. 136.—Tangle: c, chains to which are attached bun-
dles of hemp: a, iron bar to which the chains are
fastened; b, iron rings to keep the bar a off the
bottom.

manage. It does not scrape up much of the bottom like the
dredge, but catches the
shrimps, fishes and other
lively animals that usually
escape the dredge, and also
picks up sponges, echino-
derms and loose stones that
may lie in its way, often
an inconvenient weight of
the latter. On rocky or
rough bottoms it cannot be
used without danger of tearing and losing its contents.
The net should be made
with an open bottom that
can be tied up with a string

Fig. 137.—Diagram of boat dredging.

and opened again when it is hauled up for letting out the

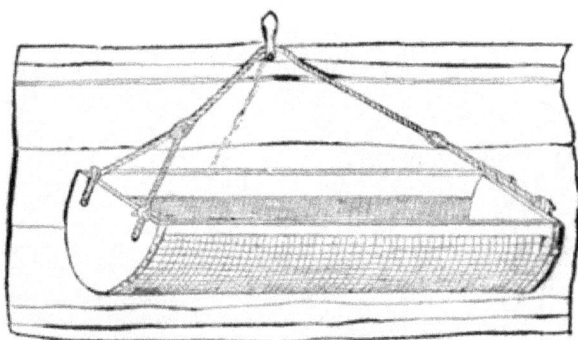

Fig. 138.—Sieve hung over a boat's side.

contents. There is another dredging instrument called a
"tangle" which can be used on any kind of bottom, Fig. 136.
It consists of bunches of twine or untwisted rope fastened

S

to chains so that they can be dragged over the bottom. Starfishes, sea-eggs, shrimps and many other animals stick to the hemp and are drawn up with it. So tightly do they hold that it is difficult to detach them after the tangle is drawn into the boat. Sometimes it will be so full of *Caprellas* (see Fig. 47, p. 46) that the color of the bunches of hemp cannot be

Fig. 139.—*Yoldia limatula.*

seen through them and sometimes sea-eggs will cover it in the same way. These are the most useful pieces of apparatus for dredging, but there are others for particular kinds of bottom. For soft and smooth mud a dredge has been made with flat thin scrapers and a frame behind to keep them level, and the bag extended so that it will slip along easily without digging up mud and catch the things lying on the sur- face. Another dredge has been made to catch the worms which are buried in the mud. This has a sharp

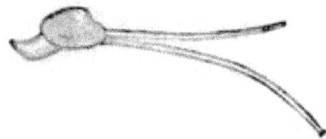

Fig. 140.—*Angulus tener*, with foot and tubes extended. Natural size.

toothed rake in front of the bag which stirs up the mud and lets it wash into it.

Dredging from a row-boat is a very simple operation ; it is only necessary to put the dredge over, pay out about twice as much rope as will reach to the bottom, make it fast to the stern and row away slowly.

Dredging from a sail-boat is a more difficult matter and needs some person accustomed to the management of boats to do it properly. If the current is strong the dredge may be fastened to the bows and the boat allowed to drift slowly

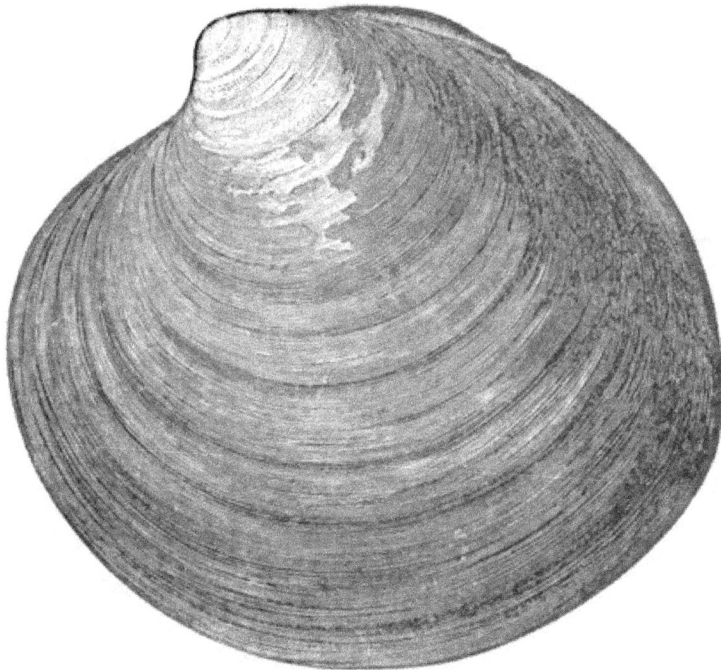

Fig. 141.—*Cyprina islandica.*

as though she were dragging an anchor ; or, if the current is not strong enough to move the dredge, a line may be put out from the stern and fastened to the dredge rope so that the boat can be pulled round with its side to the current. If the currents are not strong enough to interfere it is better to tow the dredge behind the boat as slowly as the wind will carry her.   Supposing the wind to blow in the direction of the arrow, Fig. 137, the dredge is put out on the left side, and the boat turned toward the wind until it just takes enough to keep the sails full, and it will move slowly toward the right, drawing the dredge after it.   It is best to put out the dredge while the

boat is going at a good speed, and to pay out the line slowly, so as to keep the dredge net drawn out tightly all the way down and prevent its rolling up or getting over one of the scrapers.   By holding on the rope a practised hand can tell how the dredge is going and over what kind of bottom.   If it is rocky the dredge jumps along from one stone to another going smoothly for a few moments, then catching on some corner and freeing itself again, until at length it usually becomes caught fast and stops the boat.   Then the rope must be hauled in, and when it is nearly up and down the dredge usually loosens and comes up easily without much in it.   The moving of the dredge over small stones and shells can be very dis-

Fig. 142. — Front seg-
ments of *Trophonia
affinis*, enlarged.

tinctly felt by the rope and so can the scraping into a soft or gravelly bottom ; but the motions on a smooth bottom are not so easy to understand, for the dredge will sometimes go along, wrong side out or rolled up in seaweed, in the smoothest and most promising way.

In letting down the trawl it is still more necessary to keep the rope tight so as to prevent the long net from getting foul, or the frame turning over before reaching the bottom.

When the ground has been gone over or the dredge appears to be full it is drawn up. In shallow water two or three men can do this easily by hand, but it is well to have a windlass on board in case the dredge should be caught or get filled with stones.

It is best to have a large wooden tray, in which to pick over

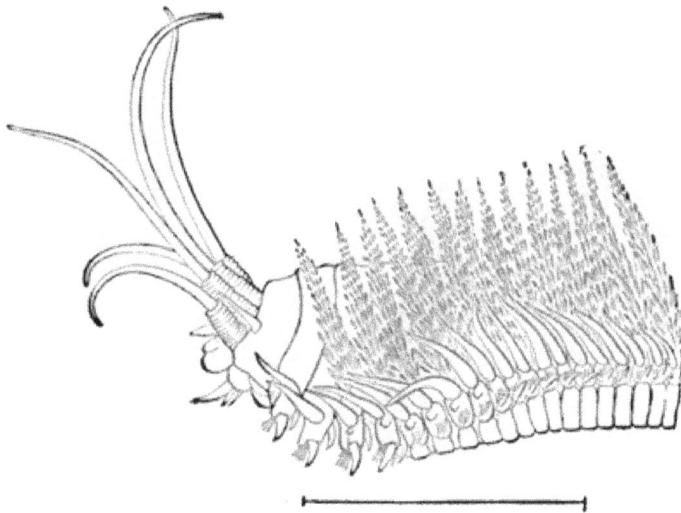

Fig. 143.—Head and front segments of *Diopatra cuprea.*

the contents of the dredge so as to keep the boat clean and to have the dirt in a convenient place to handle. If one end of the tray is open and kept over the side of the boat, the stones and mud can be washed in the tray and the water allowed to run off and the whole can be thrown overboard when done with. The tray can be made of a convenient size to lay across a small boat when dredging near the shore. If

stones and gravel come up there is nothing to do but pick
them over as carefully as possible, but if there is mud or sand
it needs to be washed in order to expose the animals. This
can be done best in a sieve made of wire gauze with ten or
twelve holes to the inch supported by a stouter wire net un-

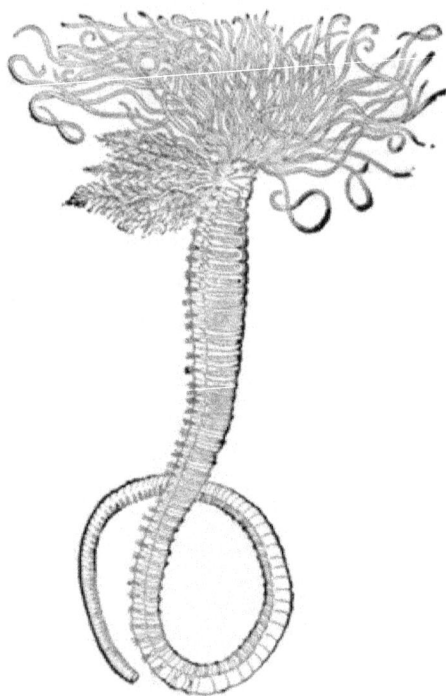

Fig. 144. — *Amphitrite ornata.*

der it. The sieve may be made flat supported by cleats so
that the water will run under it, and the washing done in the
picking-over tray ; or it may be in the shape of a half cylinder
and hung over the side of the boat, Fig. 138. This shape lets
the water through more rapidly   The most troublesome material to wash is fine mud which sticks together so closely that

the water will not go through it, and it must be stirred about with the hands or something soft until it is gradually worked through. If it is desired to examine the fine mud it should be passed through several sieves of different degrees of fineness and the finest part allowed to settle in water. The different sizes can then be picked over separately.

For a dredging boat, one that is built for fishing, with plenty of room to work on deck, is the best; and a fisherman, who is accustomed to sail it and knows the bottom, the best kind of help. A yacht is generally inconvenient and the crew over careful about the scratching of paint and spilling of dirt, which will surely happen with the greatest care. A schooner is generally considered the best, but a boat of any rig will answer the purpose if sailed by one who knows how to manage it. It is well to carry a chart and to mark on it the position of the different dredgings, as nearly as possible, and also the depth measured by a sounding line and the character of the bottom if these are not already correctly printed on the chart. A rope suitable for the dredge can not always be found on the boat and it is necessary to have one for this purpose. A small rope is, of course, more convenient to handle and take care of, but in dredging especially with a large boat, it is liable to sudden strains and one three-quarters of an inch in diameter is none too large. The kind known to the trade as "bolt-rope" is made of better material and not so tightly twisted as ordinary rope and so easier to handle and less liable to kink. The length of the rope should be at least twice the depth of the water where it is likely to be used. A number of pails should be provided beforehand for taking care of the objects dredged. There are some made

of paper which do not shrink when dry and are lighter and
cleaner than wooden pails but not so strong. Almost every
boat has a stout bucket with a rope handle for drawing water
and this is better for the purpose than the ordinary water
pail which is liable to come off the handle if put overboard.

Fig. 145.—
*Trachy-
dermon
ruber.*

For smaller objects wide-mouthed bottles of various
sizes are needed and they should be kept in boxes
or flat-bottomed baskets where they will stand up.
The common glass preserve jars are good for this
purpose and are inexpensive. The animals which
are to be kept alive should be put in clean water
and kept cool, if the weather is warm, in the
shade or if possible on ice ; but there are some things that
can not be kept alive, and even die in the dredge, and such

should be put into alcohol at
once. Specimens from different
hauls on different kinds of bottoms should not be mixed until
there is time to look them all
over.

The use of the dredge begins
where that of the hand-net ends,
beyond the rockweed and eelgrass
that cover the rocks and mud just
below low tide. It is not of much

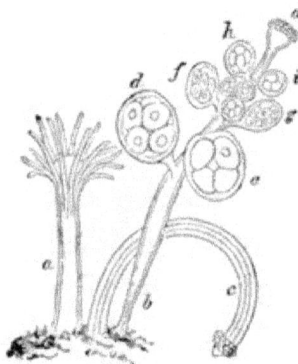

Fig. 146.—*Hydractinia poly
clina.* Female cluster.

use to dredge in the eelgrass, as the same animals are found
where they can be reached by a net, and a dredge either
slips over the grass or becomes rolled up in it, in either case
bringing up little.

Beyond the grass in harbors is usually mud filled with decaying matter that gives it the well known smell of docks and mud flats. As it extends outward, however, it becomes gradually cleaner and the animals change to those of clear water. The common clam may be found here almost anywhere and *Macoma fusca*, a small round clam already mentioned (page 23). The round clam may be also dredged near the shore. A little farther out occurs *Yoldia limatula* (page 114), a flat clam with a smooth yellow shell, often black toward the edge in large specimens; placed in water it runs out its two tubes an inch or more, and will usually kick out its curious foot by which it jumps about the dish and buries itself in the mud. The foot is nearly as long as the shell and spreads out at the end half an inch wide giving a good hold on the mud into which it may be pushed. Another pretty bivalve from the mud is *Angulus tener* (page 114). It has a large pointed foot and very long tubes which are transparent and separate to the shell. The *Cyprina islandica* (page 115), a deep-water

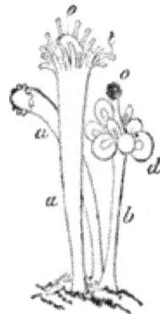

Fig. 147. — *Hydractinia polyclina.* Male cluster.

round clam, may be found in the harbor mud, especially small ones. Among the worms may be found *Clymenella torquata* (page 25), which manages to pick out sand enough to make its tubes and *Trophonia affinis* (page 116), a worm with long bristles extending forward from the head among which six soft appendages are thrust out after it has remained a short time in water. The appendages on the segments behind the head are all short and the body tapers toward the tail. This is often dug at low water on muddy shores, but small specimens have been found in considerable numbers swimming at the surface in Beverly harbor. *Nereis virens*, the common bait worm (page 23), *Nephthys ingens* and *Nephthys cæca*, are also common mud worms ; the latter have short appendages and a short proboscis with a circle of papillæ around the end that is thrust out when they are put in alcohol.

Where the mud is hard, *Diopatra cuprea* (page 117), may be found, a large and showy worm with a pair of spiral gills filled with red blood on each segment. It makes a strong tube which extends above the sand two or three inches and is there covered half an inch thick with pieces of stone and shell. It draws down below the surface quickly when touched and the tubes are oftener dredged than the worms. It lives, however, above low water in some places and can be got better with a spade. The conical shells of *Cistenides Gouldii* are often found in sand and mud sometimes with the worm enclosed.

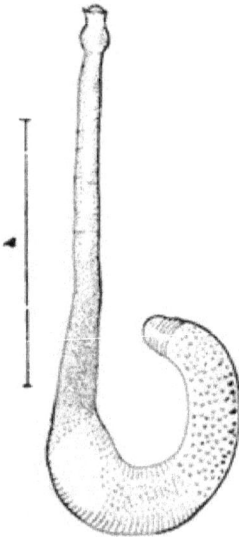

Fig. 148.—*Phascolosoma cementarium.* Enlarged.

The snails are chiefly those which can be found at low water, among them *Lunatia heros* (page 48), and its near relative, *Neverita duplicata*, whose shell is not so round and has a thickened brown piece turning out of the mouth around the middle of it. The young of *Lunatia heros* are often marked with three rows of spots running around the shell, sometimes inside as well as outside. *Illyanassa obsoleta* (page 50) is another mud snail and where it is a little cleaner, *Tritia trivittata* (page 50), one of the most active snails of the shore.

Where the bottom is hard enough to hold it grows the "devil's apron," *Laminaria*, a brown seaweed with a round stalk like India rubber and a flat blade three or four feet long and six or eight inches wide. It is attached to stones and shells by small clinging branches from the base, and among these hide away a great many animals most of them the same as found at low water on gravelly shores. Among them are the scaly worms, the long-armed starfishes, and several mud worms, *Polycirrus*, described on page 25 and *Amphitrite* (page 118) a large flesh-colored worm with long soft appendages at the head, and the rest of the body enclosed in a soft tube of mud. The *Laminaria* is often attached to a large mussel and brings it up with it, sometimes the common species and sometimes the red mussel, *Modiola modiolus*. The shells

Fig. 149.—Shell of *Buccinum undatum*.

seem to be held tightly together by the seaweed but the mussels are usually alive. The dredge often gets into this Laminaria by mistake and comes up covered with it, but if it is dredged for purposely, a grappling hook of some kind is better. After cutting off the roots the rest of the Laminaria should be looked over, before being thrown away, for polyzoa and hydroids which often cover large surfaces upon it. At very low tides the Laminaria can be got at without dredging and there are always chances of pulling it up on fish-hooks and anchors and finding it on open shores after storms.

The dredging on rocky bottoms near the shore is cleaner work, but it needs more care in the management of the boat and wears more on the dredge and rope. Comparatively little is taken up from rocky bottoms, for the more active animals hide in the crevices, and only those attached to the upper surface or among the weeds are in the way of the dredge.

The motion of the dredge over rocks can be felt by holding the rope, or even at a distance from it by the jarring of the boat. A strong dredge with a sound canvas cover over the net should be used, with a rope strong enough to hold in case of a sudden strain. On such ground the dredge jumps along picking up a little here and there and finally gets caught among the rocks and will only let go when the boat is hauled back over it, or sometimes the small line breaks and it comes up sidewise. The dredge will generally be full of red seaweeds covered round the base with polyzoa consisting largely of *Membranipora pilosa* and *Crisia eburnea* (page 60), but mixed with it are many other beautiful species which need more careful examination to tell them apart than can be given between the hauls of the dredge, so it is well to cut off the incrusted parts of the weeds and keep them in clean water till they can be looked over. There are some polyzoa to be found here that have soft shells and grow round the stems of rockweed and Irish moss.

The best specimens from rocky bottoms come up with loose stones that will almost always be found here and there, and shells of Mussels and *Cyprina mactra*. Few living mollusks expose themselves where they can get into the dredge ; but *Crepidula fornicata* (page 19), and *Crucibulum*

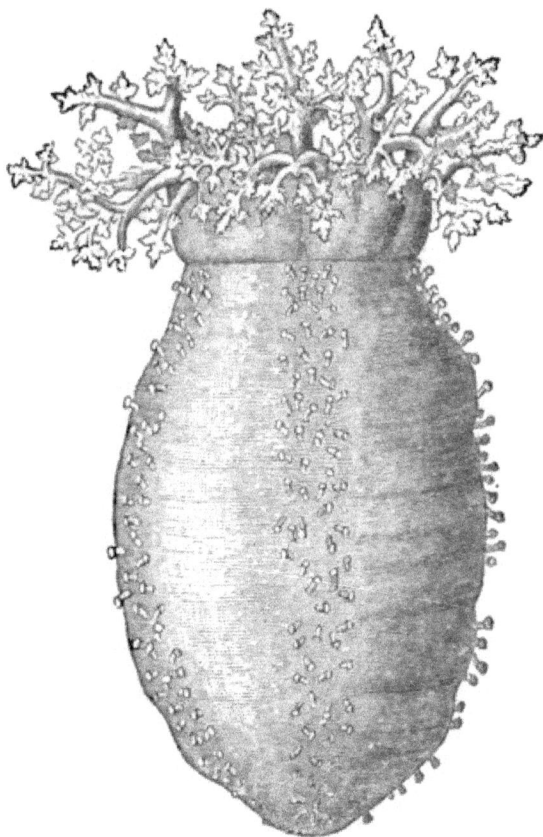

Fig. 150.— *Pentacta frondosa*, partly expanded.

striatum, a similar shell but smaller and with grooves from . the apex to the edge, come up attached to stones.

Several Chitons are also common on stones. These are snails with thick skin on the back in which are eight shelly plates lapping over each other like fish scales. The head and foot are concealed beneath and the gills are in two grooves on each side just over the foot. They cling closely to stones and resemble them in color. *Leptochiton apiculatus*, which is common south of Cape Cod, is dirty white like the stones among which it lives, while *Trachydermon ruber*, Fig. 145, a common species farther north is bright red, and lives on stones incrusted with "nullipores" of the same color. This nullipore is a plant as hard as stone that covers the stones and shells in certain places, and even the backs of crabs and lobsters with a reddish coating usually flat and smooth but rising in places into ridges and knobs. The spores grow in little hollows in the surface. *Trachydermon*

 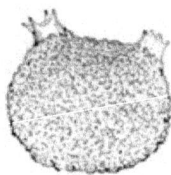

Eugyra   *Molgula*

Fig. 151.—Ascidians covered with sand.

ruber and *Chiton marmoreus* hide themselves readily on this crust, and so does another nearly white species, *Trachydermon albus.*

Another Chiton, *Amicula Emersonii*, lives in the same localities and is dirty brown like the bare stones and has branching hairs on the back. The shells of this species are small and nearly covered up by the skin.

The rocky bottoms are favorite places for the naked mollusks. *Doto coronata*, a small species with club-shaped red papillæ dotted with black, is very common and so are the green *Polycera Lessonii* and *Elysiella catulus*. The former is half an inch to an inch in length, with bright, yellow spots on the ends of the tubercles and yellow tipped gills; the latter half as large, and dark green with sometimes white spots.

Among the red seaweed lives a long-legged red crab, *Hyas coarctatus*, shaped something like the spider crab (page 44), but not more than an inch or two long and red colored. Its motions are very slow, and its back and legs are covered with seaweeds which grow there as long as the crab, so that it is hard to find. *Hippolyte Grœnlandica*, a shrimp two or three inches long is also bright red like the seaweed. Several other shrimps are also found and a *Caprella* (page 46), colored like red seaweed.

Some of the stones often have white tubes attached to them which belong to worms of the genus Serpula. These have a wreath of feather-shaped tentacles or branchiæ around the head which are expanded when the worm is at rest. A *Spirorbis* with a more delicate shell than the common shore species in an open spiral also occurs on the seaweed.

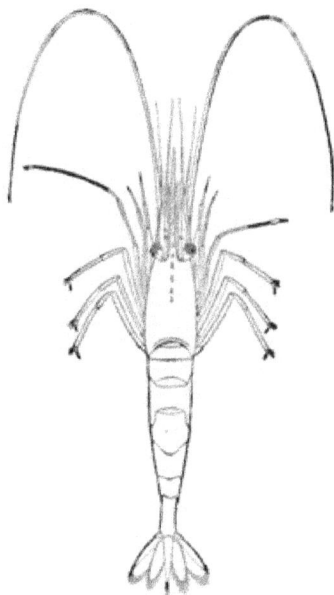

Fig. 152.—*Pandalus annulicornis.*

The shore starfishes and especially the sea-eggs are very common on rocky bottoms and it is not uncommon to get a dredge full of the latter mixed with stones.

The common sea-anemone and another kind, the thick-armed sea-anemone, *Urticina crassicornis*, are both found on stones ; the latter has fewer and thicker tentacles and is red, or marked with red spots, around the sides and bases of the tentacles. They sometimes contract so as to be almost flat.

In some places the whole bottom is covered with small stones and coarse gravel which come up in large quantities in the dredge, bringing with them the same animals which escape when attached to larger stones. Sometimes immense numbers of shells of dead bivalves are mixed with the stones or even form the whole covering of the bottom. Shells of *Mytilus* and *Modiola* often occur in this way, and also those of *Cyprina islandica* and *Mactra solidissima* in places where a living specimen of either species is seldom found ; for they hold on or dig into the bottom when alive and so are passed over by the dredge.

These shelly bottoms are favorite places for hermit crabs (page 45), which inhabit the dead snails' shells of all kinds, especially those of *Lunatia heros* of various sizes. The shells inhabited by crabs are often covered with a growth of white or pink animals, *Hydractinia polyclina*, Figs. 146 and 147, which should be carefully examined. They form a crust all over the shell from which extend upward little polyps something like *Clava*, but of several different kinds in different parts of the colony. In some spots most of the individuals are females with ovaries along the sides filled with eggs and mixed with them are other sterile ones. In other places a large part are males mixed with sterile ones of a different kind.

Many of the smaller shells are occupied in another way by a "Sipunculoid worm" *Phascolosoma cementarium*, Fig. 148, a common species on the northern coasts of New England. "This worm takes possession of a dead shell of some small Gasteropod, like the hermit-crab, but as the aperture is always too large for the passage of its body, it fills up the space around it with a very hard and durable cement, composed of mud and sand united together by a secretion from the animal· leaving only a small, round opening, through which the worm

can extend the anterior part of its body to the distance of one or two inches, and into which it can entirely withdraw at will. It thus lives permanently in its borrowed shell, dragging it about wherever it wishes to go, by the powerful contractions of its body, which can be extended in all directions and is very changeable in form. When fully extended the forward or retractile part is long and slender, and furnished close to the end with a circle of small, slender tentacles, which surround the mouth; there is a band of minute spinules just back of the tentacles; the anal orifice

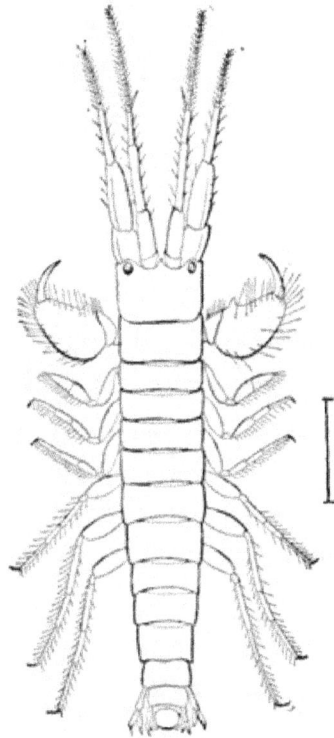

Unciola irrorata Say.

Fig. 153.—*Unciola irrorata.*

is at the base of the retractile part; the region posterior to this has a firmer and more granulous skin, and is furnished toward the posterior end with a broad band of recurved spines, which evidently aid it in retaining its position in the shell. As it grows too large for its habitation, instead of changing it for a larger shell as the hermit-crabs do, it gradually extends its tube outward beyond the aperture by adding new materials to it.

A larger sipunculoid, *Phascolosoma Gouldii*, lives in gravelly bottoms near low water where it is hard digging.

The dead shells are usually full of holes running along under the surface, so that they are completely honeycombed

Fig. 154. — *Margarita obscura.*

and break in pieces easily. These holes are inhabited by worms which can not easily be pulled out, but creep out themselves after they have been standing for some time without the water being changed. Some of these holes in shells are caused by sponges which fill them up and extend little finger-shaped processes through the surface of the shell.

On the southern coast of New England two of the largest snails on the coast are found on these bottoms, *Sycotypus canaliculatus* and *Fulgur carica*, with shells six or eight inches long.

Crepidula fornicata is very common and sometimes the bottom is covered with its dead shells.

Fig. 155. — *Astarte sulcata.*

Buccinum undatum, Fig. 149, is a common shell farther north, up to low water. This is the English "Whelk," and in Europe is much used as an article of food.

On both rocks and shells are often found white or variously colored masses looking like pork or tapioca pudding. These are compound ascidians of the genera *Amaroecium* and *Leptoclinum.* The individuals are very small and connected by a thick jelly-like substance through which are the openings by which the water passes in and out. The structure is best seen by cutting a slice.

With the starfishes and echini there occurs, south of Cape Cod, a dark purple Holothurian, *Thyone Briareus*, four or five inches long, covered with fine papillæ all over the surface, and in Massachusetts bay and farther north, *Pentacta frondosa*, Fig. 150, the largest Holothurian of the coast. This is dark brown, five or six inches long and shaped like a cucumber, with five grooves from one end to the other in which are five rows of suckers. When expanded in water it puts out a circle of ten branching tentacles and may be preserved partly expanded by tying a string around below them before it has time to draw them in.

Another Holothurian, *Caudina arenata*, lives buried in sand, but is sometimes washed ashore after storms. It is shaped quite differently from *Pentacta*, one end of the body being drawn out narrow so that it can be extended up to the surface while the rest is buried. The mouth is at the end of this narrow portion.

On sandy bottoms the "sand-dollar" is sometimes dredged in great quantities, as are also several Ascidians that cover themselves with a coating of sand which adheres even after they are taken out of the water, Fig. 151. While they are contracted they are entirely concealed and look like balls of sand ; but when at rest they put out two tubes, at the ends of which are the holes through which the water runs in and out.

The animals which have been mentioned thus far are the ones most likely to be met with along the northern coast in bays and harbors, and we will now only mention a few which

live in colder and deeper water, at different depths in different places, on hard, muddy, or sandy bottoms.

The common shrimp, *Crangon vulgaris* (page 36), is found
from low water down to a considerable depth, where another
stouter species, *Crangon boreas,* occurs with a more spiny
shell and marked with spots of reddish-brown. The most
common deep water shrimp is *Pandalus annulicornis,* Fig.
152; it often gets into the
dredge and still oftener into the
trawl, and usually dies before
reaching the surface. It is
partly transparent and marked
with red on the edges of the
segments. Another species,
Pandalus borealis, grows larger,
five or six inches long, and
lives in deeper water. Both
species are good eating, but
the front part of the body containing the stomach should be
thrown away. With these are
several other shrimps of the
genus *Hippolyte,* most of them
beautifully marked with red.

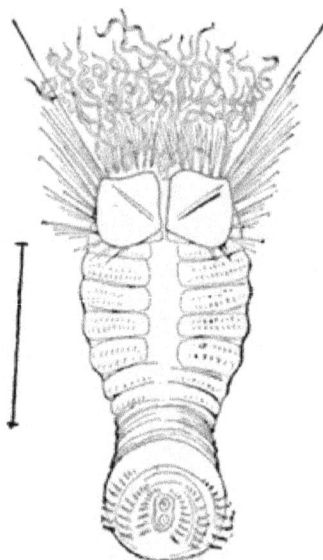

Fig. 156.—*Sternaspis fossor.*

There are two Amphipods which are commonly found, in
dredging, on various kinds of bottom and even at low water,
and might have been mentioned before. One is *Unciola
irrorata,* Fig. 153. It is covered with red and white markings when fresh, and is much more flattened up and down

than most of the group, so that it can rest back up. The other is a stout species, *Ptilocheirus pinguis*, which is gray or purplish with a lighter border round the segments.

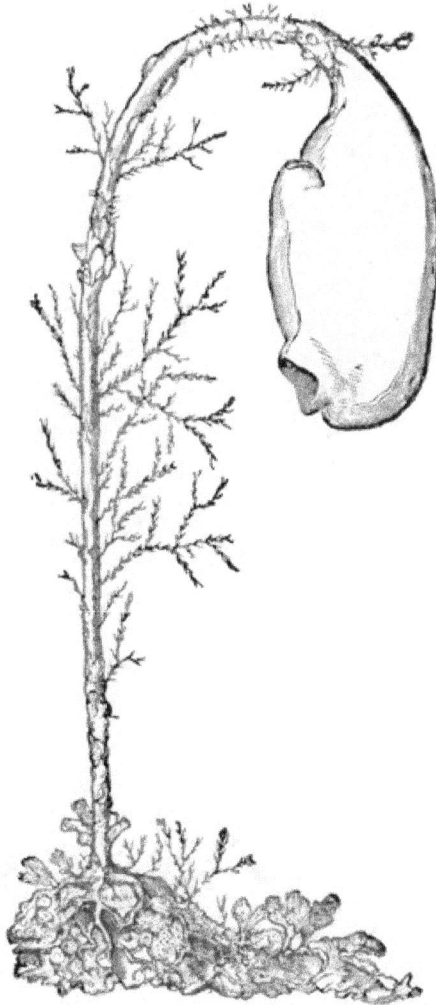

Fig. 157. — *Boltenia Bolteni*, a long-stalked Ascidian.

Among the snails besides *Lunatia heros*, a much smaller

species, *Lunatia immaculata*, is common.   It is white, smoother and more pointed than the large species.

Buccinum undatum (page 123), is still found here.   *Fusus decemcostatus* is a large shell two or three inches long with high ridges around it.   *Margarita obscura*, Fig. 154, and other species of Margarita are short shells with ridges running around them and they have a pearly color.

Menestho albula is a long whitish shell about an inch long. *Scalaria Grænlandica* is an inch to an inch and a half long, and has raised ridges running across it at short distances apart.

Fig. 158. — *Terebratulina septentrionalis.*

There are several naked snails here that are not found near the shore.   One of these is *Dendronotus robustus*, a stouter species than *arborescens* (page 52), with a wider head and shorter gills.   There are also several species of *Eolis*.

Among the bivalves are *Astarte sulcata*, Fig. 155, *Astarte castanea*, a smooth species, and *Astarte quadrans*.

Nucula delphinodonta is a little shell that looks like a grain of gravel and is often found mixed with it.   *Ledo tenuisulcata* is a thin shining shell like *Yoldia limatula* (page 114), but smaller and with one end long and narrow.

There are two very conspicuous worms, one a small worm

Sternaspis fossor, Fig. 156 (much enlarged), with a pair of little shells on its back. The other is one of the largest worms, *Aphrodite aculeata*, which grows to be four or five inches long and two inches wide. It is related to the scaly worms (page 26), and is covered with long bristles of various shapes that curve up over the back and make it look like a small quadruped covered with fur. Small specimens are oftener found than those of the largest size.

The most conspicuous of the deeper water ascidians is *Boltenia Bolteni*, Fig. 157, which gets to be two or three inches long and has a stem six or eight inches long by which it is attached to the bottom. Small individuals are often found on seaweed. Another large Ascidian is *Cynthia pyriformis*, which becomes even larger, but is attached by a wide base to the bottom. It is sometimes brightly colored red and yellow like a peach.

The common *Terebratulina septentrionalis*, Fig. 158, is attached to stones or lumps of mud by a short stem. It has

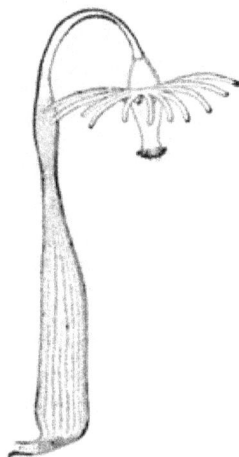

Fig. 159. — *Corymorpha pendula*. The thick end of the stem extends below the sand.

an upper and under shell which open a short distance when alive and show the two spiral arms or gills at the sides of the mouth. This does not belong with the bivalve mollusks but to the Brachiopods, a class of animals of which there are very few living species, but great numbers of fossils. The shells are often covered by a sponge which grows larger than the shell itself.

On the same bottom with *Terebratala* grow many sponges. *Grantia ciliata* is an egg-shaped sponge with an opening in the upper end surrounded by long needles.

A species of *Polymastia* grows over stones or shells and has long papillæ extending up from various parts of it which have their spicules arranged in two sets, one running lengthwise and the other around so as to form nearly square openings. These sometimes get torn off and come up alone in the dredge. A species of *Tethya* grows in flat masses covered

Fig. 160. — *Alcyonium carneum*. Three polyps, enlarged four times.

with long fine spicules like hairs. Other sponges grow in large soft yellow masses that fall to pieces soon after leaving the water.

Where the bottom is sandy, one of the commonest animals is *Corymorpha pendula*, Fig. 159. It is colored pink and white and grows six inches long. The dilated end of the stem is embedded in the sand.

There is a compound polyp, *Alcyonium carneum*, Fig. 160, that grows in large flesh colored clusters attached to stones. These clusters are entirely soft and make no coral.

On this bottom are several other Echinoderms. The *Lopho-thuria Fabricii*, Fig. 161, is one of the commonest. While Pentacta has a soft body with fine rows of suckers all alike,

Fig. 161. — *Lophothuria Fabricii* expanded, turned up so as to show the soft under side.

the Lophothuria has only one side soft and the three rows of suckers belonging to it. The other side is covered with hard scales of a bright red color. The animal rests and crawls on the soft side so that it does not look at all like a radiate animal. The tentacles round the mouth are usually drawn in when they come up, but after resting a short time in water they will expand like those of Pentacta.

The same starfishes as are on the shore are found down to considerable depths and with them a more delicate white one. *Leptasterias tenera* and a ten-armed one, *Solaster endeca*.

Great numbers of interesting animals are brought up by fishermen in lobster-traps, seines and even on fish-hooks, especially where the latter are left down a long time as on "trawls ;" but it is hard to induce these men to bring in anything except marketable fishes even for a fair price. When this can be done, however, a surprising quantity of new animals is sure to be found, as has been the case at Gloucester during the past few years where fishermen have become interested in the work of the U. S. Fish Commission and have brought in cartloads of things that were formerly thrown away and were almost unknown to naturalists.

Other animals, especially shells, are to be found in the stomachs of fishes which have swallowed them in deep water, and great numbers of interesting things have been collected from this source.

INDEX.

INDEX.

	PAGE
Actinia	67
Alcyonium carneum	136
Amarœcium	130
Amicula Emersonii	126
Amphipods	28, 44
Amphitrite	123
Angulus tener	121
Animals between tides	1–10
Animals below tides	33
Anomia	19
Aphrodite aculeata	135
Ascidians	56–130, 135
Ascidians, compound	130
Asterias	59
Astrangia	70
Astarte castanea	134
Astarte quadrans	134
Astarte sulcata	134
Autolytus	55, 86
Aurelia	99
Barnacles	11
Beach animals	11–32
Beach snails	16
Blue crab	41
Bolina	107
Boltenia bolteni	135
Bottom animals	109
Brachiopods	135
Buccinum undatum	130, 134
Burrowing worms	23
Caligus	85
Callinectes	41
Cancer	38
Caprella	44
Caudina arenata	131
Cephalopods	89

	PAGE
Cerebratulus	27
Chiton marmoreus	126
Chitons	125
Cistenides Gouldii	24, 122
Common clam	19, 20, 21
Clava	73
Clione papillionacea	94
Clymenella torquata	24, 25
Copepods	82
Coral	70
Corymorpha pendula	136
Coryne mirabilis	74, 125
Crabs	38
Crangon	132
Crepidula fornicata	18, 125, 130
Cribrella	98
Crisia eburnea	124
Ctenophoræ	106
Cumacea	86
Cuttlefishes	89
Cyanea	101
Cynthia pyriformis	135
Cyprina islandica	121
Dendronotus arborescens	48, 134
Dendronotus robustus	134
Devil's apron	16
Diopatra cuprea	122
Doto coronata	48, 126
Dredges	109
Echinus	63
Elysiella catulus	126
Eolis	48, 134
Fabricia Leidyi	55
Fiddler crabs	30
Fish eggs	88

	PAGE
Fulgur carica	130
Fusus decemcostatus	134
Gammarus	28
Gelasimus	30
Hen clam	21
Hermit crabs	43
Hippolyte grœnlandica	121
Holothurians	66, 131
Horseshoe crab	44
Hyas coarctatus	127
Hybocodon	105
Hydractinia	128
Hydroids	74
Hyperia	86
Idotæa	29, 86
Idyia	107
Infusoria	78
Isopods	29, 85
Jelly-fishes	98
Lacuna vincta	46
Laminaria	16, 123
Ledo tenuisulcata	134
Lepidonotus	26
Leptasterias tenera	138
Leptochiton apiculatus	126
Leptoclinum	130
Leptoplana variabilis	27
Leptosynapta Girardii	66
Limnoria	39
Limpet	18
Limulus	44
Lineus viridis	26
Littorina	17, 18
Littorina litorea	17
Littorina palliata	17
Littorina rudis	17
Lobster	34
Lophothuria	98
Lophothuria Fabricii	137
Lunatia heros	17, 45, 132, 133
Lunatia immaculata	134
Macoma fusca	22, 121
Mactra solidissima	21
Margarita obscura	134

	PAGE
Metridium marginatum	67
Megalops	39
Membranipora pilosa	124
Menestho albula	134
Modiola modiolus	16, 123
Modiola plicatula	15
Molgula	56
Mud worms	23
Mussels	14
Mya arenaria	19
Mysis	85
Mytilus edulis	14
Naked snails	47
Nanomia	105
Nassa obsoleta	46, 122
Nemertines	26
Nephthys	122
Nereis	23, 86, 122
Nereis virens	23, 186
Neverita duplicata	122
Nicolea simplex	54
Nucula delphinodonta	134
Nudibranchs	47, 126
Obelia	102
Oceania languida	104
Ophiurans	62
Oyster	49
Oyster crab	41
Pandalus	132
Panopæus	40
Pecten	51
Pedicellariæ	64
Pentacta frondosa	125, 131–137, 138
Periwinkle	17
Phascolosoma cementarium	128
Phascolosoma Gouldii	130
Physalia	106
Planarians	27
Platyonichus	41
Pleurobrachia	107
Polycera Lessonii	126
Polycirrus	25, 123
Polymastra	136
Polyp, compound	136
Polyzoa	58, 124
Portuguese man-of-war	106
Pteropods	94

	PAGE
Ptilocheirus pinguis . .	133
Purpura lapillus . . .	16
Purpura lapillus, eggs of, .	17
Pycnogonidæ . . .	45
Quahaug	21
Razor clam	20
Round clam	21
Sagartia leucolena . .	69
Sagitta	88
Salaria Grœnlandica . .	134
Salpa	96
Sand dollar . . . 66,	131
Saxicava . . .	20
Scaly worms . . .	26
Scollop	51
Sea anemone . . .	67
Sea egg . . .	63
Ship worm . . .	52
Shore animals . . .	11
Shrimp . . . 37,	132
Siliqua costata . . .	22
Solaster endeca . . .	138
Spider crab . . .	42
Sponges . . 75, 135,	136
Squids	89
Starfishes . . .	59
Sternaspis fossor . .	135

	PAGE
Surface animals . . .	81
Sycotypus canaliculatus .	130
Tangle	113
Tectura testudinalis . .	18
Terebratulina septentrionalis	135
Teredo	52
Tethya	136
Thyone briareus . . .	131
Tiaropsis	104
Tima formosa . . .	104
Trachydermon ruber . .	126
Trawl	112
Tritia trivittata . . 46,	122
Trophonia affinis . . .	121
Tubularia . . .	72
Unciola irrorata . . .	32
Urosalpinx cinerea . .	17
Urticina crassicornis . .	127
Venus mercenarea . .	21
Worms . . . 23,	54
Worm tubes . . .	24
Yoldia limatula . . 121–134
Zoea	38